Dr Hairy's Research Summaries

By Dr David Hindmarsh
and Edward Picot

ISBN 978-1-312-22722-4

Self-published via Lulu (www.lulu.com)

Table of Contents

QUIZ NO 11

QUIZ NO 12

Introduction

A bit about Dr Hairy and this book

The Dr Hairy project is an attempt to produce humorous and down-to-earth training materials, written in an entertaining and non-academic style, for GPs and Registrars. It is the brainchild of Dr David Hindmarsh, a GP from Kent, and Edward Picot, his Practice Manager, who also happens to be a writer and new media/video artist.

The first Dr Hairy book was published in 2007: *Dr Hairy's Guide to the GP Curriculum*.

The Dr Hairy videos (puppet-animations about the adventures and misadventures of a fictional GP called Dr Hairy) started to appear online in 2010.

In 2011 we started to run Dr Hairy's Professional Development Course, using the Dr Hairy videos as material for reflective learning, and a book based on this course was published by Scion Publishing in 2012 - *Professional Development for Appraisal and Revalidation: The Dr Hairy Workbook,* which won the Primary Care section of the 2012 BMA Book Awards.

Also in 2012, we set up the Dr Hairy website (http://drhairy.org) and self-published a pack of course-materials entitled *Dr Hairy's Professional Development Course,* plus a stand-alone DVD of the Dr Hairy videos, entitled *Dr Hairy in...*

In 2013 David came up with the idea of summarising research articles in the Dr Hairy style. After a bit of debate we decided to bring these out monthly in bundles of about twelve, with a true-or-false quiz at the end of each bundle, not so much to test people's knowledge as to help fix in their minds what they had just read. We offered these on our website under the collective title *Dr Hairy's Research Summaries,* and having now completed a whole year's worth we have put them together in the form of the book you are reading.

True to the spirit of the Dr Hairy project, we have tried to keep the summaries as plainly-written and concise as possible, and we have also popped in a few jokes here and there to liven up the proceedings. If you want to refer to the original articles, full references and the names of the authors are given at the beginning of each summary.

Of course, new research is appearing all the time, and if you read through these summaries in 2020 you will probably find that much of the material here looks a bit out of date. On the other hand there are certain recurring themes here which may remain interesting for quite a long time – for example the government's vigorous promotion of a preventive healthcare agenda, often via Primary Care, often to the detriment of Secondary Care, often on the basis of rather scant evidence, and often greatly to the benefit of the pharmaceutical companies.

And hopefully, even in 2020, some of the jokes will still be funny.

- Edward Picot and David Hindmarsh, 2014

Influence of preceding length of anticoagulant treatment and initial presentation of venous thromboembolism on risk of recurrence after stopping treatment: analysis of individual participants' data from seven trials

BMJ 2011;342:d3036

Florent Boutitie, statistical investigator, Laurent Pinede, investigator, Sam Schulman, professor, Giancarlo Agnelli, professor, Gary Raskob, professor, Jim Julian, statistical investigator, Jack Hirsh, professor emeritus, Clive Kearon, professor

- Case fatality rate is 10%
- VTE often recurs – 20-30% risk in 2-8 years
- No benefit of continuing treatment for more than 3 months
- USA guidelines – 3 months' treatment for unprovoked VTE and indefinite treatment for unprovoked proximal DVT or PE
- Bleeding case fatality rate is 9-13%
- Fatality from recurrent VTE decreases with time and fatality from bleeding increases with time

Ovarian cancer – NICE guidelines

BMJ 2011;342;d2073

C Redman, consultant gynaecological oncologist, S Duffy, medical director, N Bromham, researcher, K Francis, senior researcher on behalf of the Guideline Development Group

- The incidence is rising . It is the 5th most common cancer in women, with a lifetime risk of 2%
- Five year survival is 35%
- Refer to gynaecologist if abdominal or pelvic mass and/or ascites
- Symptoms include bloating/abdominal distension, loss of appetite, pelvic or abdominal pain, urinary frequency or urgency and should be investigated especially in women over 50 years old
- Women over 50 yo with IBS type symptoms should be investigated for ovarian cancer as IBS rarely presents for the first time at this age
- Measure CA125 in women with suggestive symptoms
- If CA 125 is > 35 , arrange US scan of abdomen and pelvis
- In women < 40 yo consider measuring AFP and HCG

Otorrhoea

BMJ 2011; 342-d2299

Miran Pankhania, specialist trainee year one in otolaryngology, Owen Judd, specialist registrar in otolaryngology, Andy Ward, general practitioner

- A GP with a list size of 2000 will see one new case of cholesteatoma approximately every 5 years.
- Pain over mastoid or temporal bone on percussion suggests bony involvement
- Pseudomonus and staphylococcus aureus account for 70% of cases of otitis externa – aminoglycoside drops are a good treatment option for these organisms.
- Aminoglycoside drops are safe for up to 2 weeks use in the presence of a perforated tympanic membrane . The drops may remain in the inner ear fluid for up to 6 months so further courses should be used with caution.
- Refer to ENT if patients with otitis externa require more than 2 treatment attempts – take a swab first.
- Hearing aids should not be worn during treatment for otitis externa
- Refer immunosuppressed patients with unrelenting pain or focal neurology as they may have necrotising otitis externa or pseudomonal osteomyelitis

Epididymo-orchitis

BMJ 2011;342;d1543

Rachel Sacks, **Specialty Doctor in GUM/HIV, Linda Greene, Consultant in GUM/HIV**

- A rapidly growing testicular tumour can cause pain
- Idiopathic scrotal oedema causes bilateral oedema and scrotal bruising with no testicular tenderness
- If under 35 years old there is a high risk of STI – treat and refer – consider doxycycline 100mg bd for 14 days or ofloxacin 200mg bd for 14 days
- If there is low risk of STI ciprofloxacin 500mg bd for 14 days is advised
- First pass urine should be sent
- Scrotal swelling may take 6 weeks to settle
- If there is no improvement after 3-5 days, consider a scrotal abscess

New NICE guidelines for hypertension

BMJ 2011;343:d5644

L D Ritchie, Mackenzie professor of general practice, N C Campbell, reader in general practice, P Murchie, senior lecturer in general practice

- Target >80 = 150/90, everyone else = 140/90
- Thiazides are out but don't have to be changed if BP is well controlled
- Indapamide and chlorthalidone are diuretics of choice
- Calcium blockers are first choice for >55yo
- ACE or ARB first choice < 55
- Ambulatory monitoring offered if BP 140/90 or more
- Consider immediate treatment if BP 180/110 or more
- Ambulatory monitoring is more specific and sensitive than home or clinic monitoring (systematic review) – but... meta analysis showed home monitoring may be as good as ambulatory at predicting cardiovascular events – so it may be a suitable alternative (especially if you haven't got an ambulatory monitor, like me)

Cardiovascular safety of NSAIDs

BMJ 2011;341;d6618

Wayne A Ray, professor and director, Division of Pharmacoepidemiology

- Patients at high risk of cardiovascular disease should not use cox-2 inhibitors because of the increased risk of serious cardiovascular disease
- There is limited data from clinical trials – information is from a cohort study and from a recent network meta-analysis
- Network analysis makes various assumptions and the usefulness of this with NSAIDs is limited as too few adequately powered clinical trials exist
- Naproxen seems to be safer than diclofenac in patients at high risk of cardiovascular disease

Diagnosis and management of premenstrual disorders

BMJ 2011;342-D2994

Shaughn O'Brien, professor of academic obstetrics and gynaecology, Andrea Rapkin, professor of obstetrics and gynaecology, Lorraine Dennerstein, professorial fellow, Tracy Nevatte, postdoctoral research associate

- Symptoms are clearly related to ovulation
- Symptoms disappear by the end of menstruation
- Symptoms must be rated and may cause substantial impairment
- Non drug treatment includes CBT, relaxation, exercise, supplements – calcium carbonate, magnesium oxide, vitamin B6 100mg a day , fruit extract of vitex agnus castus.
- SSRIs and buspirone may be effective
- Progestogens are not indicated
- Danazol in luteal phase helps with mastalgia
- Yasmin is the pill of choice
- Estradiol patches and mirena is effective treatment
- Diagnosis is based on 2 months of recording and rating symptoms

The management of tennis elbow

BMJ 2011;342:d2687

John Orchard, adjunct associate professor, Alex Kountouris, physiotherapist

- Tennis elbow is an underuse-overuse tendinopathy
- Pain on resisted finger extension and full elbow movements are found
- Double-handed backhand is less likely to cause it
- The crosscourt forehand return dipping sharply at the server's feet may aggravate the condition but get you to love-fifteen
- Cortisone injections are harmful in the long term
- Eccentric lengthening exercises are effective
- GTN patches may help
- Botox may be considered before surgery but both are disabling for many months

Measles, mumps, and rubella vaccination in a child with suspected egg allergy

BMJ 2011;343;d4536

Alexandra Rolfe, clinical research fellow, Aziz Sheikh, professor of primary care research & development & honorary consultant

- Egg allergy usually presents as angioedema, urticaria, or gastrointestinal symptoms
- Egg allergy has a prevalence of 1-2% in children aged 2.5 years.
- IgE and RAST can be tested if there is a suspicion of egg allergy
- Risk with MMR vaccine is very low even in children with severe egg allergy
- The MMR vaccine is cultured on chick embryo fibroblasts and the amount of egg protein in the vaccine is negligible
- Egg allergy resolves spontaneously in the majority of cases
- Once it does resolve, have two boiled eggs for breakfast on a Saturday morning with Marmite fingers: yummy

Postural hypotension

BMJ 2011;342;d3128

T Sathyapalan, senior lecturer, M M Aye, clinical research fellow, S L Atkin, professor

- Postural hypotension is a decrease of at least 20mm Hg systolic and 10mm Hg diastolic within 3 minutes of standing upright – the heart rate may also increase, but an excessive rise(>30beats/minute)is diagnostic of postural tachycardia syndrome
- Symptoms include dizziness, fatigue, sweating, visual and hearing disturbance, arrhythmias and aggravation of angina
- Reduced sweating, gastroparesis, incontinence or urinary retention, constipation and erectile dysfunction all suggest autonomic dysfunction and symptoms may be worse with excessive nocturia or after a meal
- If on standing up suddenly you break out into a sweat, lose your erection, feel constipated and pee your pants, you should probably sit down again
- Hyponatraemia and hyperkalaemia may point towards hypocortisolaemia as a cause for postural hypotension
- Parkinson's disease, cerebrovascular disease and dementia may coexist with postural hypotension

Advising on travel during pregnancy

BMJ 2011;342;d2506

Natasha L Hezelgrave, academic clinical fellow, obstetrics and gynaecology, Christopher J M Whitty, professor, Andrew H Shennan, professor of obstetrics, Lucy C Chappell, clinical senior lecturer

- After 28 weeks women need a letter from Dr or midwife confirming the due date and that the pregnancy is uncomplicated
- Flying is permitted to 36/40 usually in singleton pregnancies and to 32 weeks in multiple pregnancies
- Cosmic radiation on a flight is greater than that on the ground and 20 hours flying is equivalent to the radiation from a chest x-ray . The risk from a single flight to the fetus is negligible but frequent flyers may exceed the recommended maximum
- RCOG advises against travel to malarial areas in pregnancy but if unavoidable, chemoprophylaxis carries low risk to the fetus especially in the 2nd and 3rd trimesters
- Live polio vaccine and killed typhoid vaccines should be avoided in pregnancy unless there is high risk of exposure
- MMR and oral typhoid are contraindicated
- Most other killed vaccines and toxoids are considered safe as is yellow fever
- Patients brought up in a malarial area may have malarial anaemia without parasites in the blood because of placental sequestration of parasites
- Babies born in mid-air may go on to become either pilots, sky-divers or human cannonballs

Giant cell arteritis

BMJ 2011;342;d3019

Nada Hassan, specialist registrar, Bhaskar Dasgupta, consultant rheumatologist, Kevin Barraclough, general practitioner

- About 20% of patients with polmyalgia have temporal arterits and about 50% of patients with temporal arteritis have symptoms of polymyalgia
- The sensitivity of individual clinical features is low
- Acute blindness occurs in 20% of patients with temporal arteritis without prompt treatment
- Jaw or tongue claudication occur in a minority of cases and herald impending ischaemia
- Jaw claudication, diplopia and abnormal palpation of temporal arteries are most predictive – likelihood ratio >2
- Likelihood ratio of >1.5 for scalp tenderness, temporal headache,ESR>100, anaemia
- Temporal artery biopsy should occur within 2 weeks of starting steroids
- Start prednisolone 40mg in patients without ischaemic symptoms
- Admit if visual symptoms
- Start 60mg prednisolone if ischaemic (claudication) symptoms
- Reduce prednisolone by 10 mg steps every 2 weeks until dose is 20mg a day , then in 2.5mg steps

Vertebroplasty for vertebral fracture

BMJ 2011;343:d3470

David J Wilson, consultant musculoskeletal radiologist

- Osteoporotic vertebral collapse may affect 1 in 4 adults over 50 years old
- Cement augmentation may improve symptoms in 70-90% of patients
- NICE guidelines advise conservative treatment for at least 4 weeks
- Two randomised controlled trials in 2009 suggest there is some doubt about the effectiveness of vertebroplasty
- There is insufficient evidence currently to reach firm conclusions about the value of vertebroplasty

Management of nausea and vomiting in pregnancy

BMJ 2011;342;d3606

Sheba Jarvis, academic specialist registrar in endocrinology and diabetes, Catherine Nelson-Piercy, professor of obstetric medicine

- Typically peaks at 9 weeks
- Hyperemesis gravidarum occurs in less than 1% of pregnancies
- Hyperemesis can cause Wernicke's encephalopathy
- Primiparous women, non-smokers , and those who are younger and of lower socio-economic class are higher risk
- Pregnancies complicated by nausea and vomiting are less likely to result in miscarriage
- Nausea is mediated by placental HCG
- Thyroid function may be altered in pregnancy because HCG and TSH are similar structures
- Relaxation of smooth muscle is mediated by progesterone causing reduced oesophageal pressure and delayed gastric emptying
- Hyperemesis tends to recur in subsequent pregnancies
- Ginger, chamomile and peppermint, acupuncture and vitamin B6 may help
- Most anti-emetics are safe in pregnancy - cyclizine 50mg tds , metoclopramide, prochlorperazine, promethazine 25mg nocte, domperidone, ondansetron 4-8mg up to tds

QUIZ NO 1

Anticoagulation for venous thromboembolism: VTE is not likely to recur in 2-8 years	☐	True
	☐	False
Unprovoked proximal DVT or pulmonary embolism requires life-long treatment	☐	True
	☐	False
Likelihood of death from anticoagulation decreases with length of treatment, but fatality from recurrent VTE increases with length of anticoagulation	☐	True
	☐	False
Likelihood of death from anticoagulation decreases dramatically after you've died	☐	True
	☐	False
Ovarian cancer: The incidence of ovarian cancer is falling	☐	True
	☐	False
Women over 50 with IBS symptoms should not be investigated for ovarian cancer because IBS (irritable bowel syndrome) often presents in over 50s	☐	True
	☐	False
CA125 should not be used as a screening test in women with suggestive symptoms	☐	True
	☐	False
Otorrhoea: An average GP will "see" a new case of cholecteatoma every 20 years or so	☐	True
	☐	False
Immunosuppressed patients with constant otalgia and neurological signs probably don't have necrotising otitis externa or osteomyelitis, and should be advised to drink strong tea, inhale steam, and listen to the slow movement of Elgar's cello concerto	☐	True
	☐	False
Epididymo-orchitis: In men under 35 years old prescribe doxycycline 100mg bd or ofloxacin 200mg bd for 7 days	☐	True
	☐	False
If there is a low risk of STI, amoxicillin is advised	☐	True
	☐	False
NICE hypertension guidelines: Thiazides should be changed in all patients	☐	True
	☐	False

Home monitoring is as good as ambulatory monitoring at predicting cardiovascular events	☐ True ☐ False
NSAIDs and cardiac risk: Naproxen is safer than diclofenac in patients at risk of cardiovascular disease	☐ True ☐ False
The data regarding NSAIDs and cardiac risk comes from adequately powered clinical trials	☐ True ☐ False
"Adequately powered" means they used Duracell batteries in their calculators	☐ True ☐ False
PMS: progestogens may help with PMS	☐ True ☐ False
Extract of vitus agnus castus may help PMS symptoms	☐ True ☐ False
Tennis Elbow: Pain on resisted finger extension is found in tennis elbow	☐ True ☐ False
Cortisone injections are effective long-term	☐ True ☐ False
English Tennis Elbow is also known as "Backhand into the net syndrome" or "Tim Henman's curse": the only cure is new balls	☐ True ☐ False
MMR and suspected egg allergy: Egg allergy usually causes anaphylaxis	☐ True ☐ False
There are large amounts of egg protein in the MMR vaccine	☐ True ☐ False
Postural hypotension: A rise in pulse rate from sitting to standing of more than 30 beats per minute is diagnostic of postural tachycardia syndrome	☐ True ☐ False
Hypokalaemia may be associated with postural hypotension due to hypocortisolaemia	☐ True ☐ False
Travel in pregnancy: It is fine to fly until 36 weeks in multiple pregnancies	☐ True ☐ False

The radiation from a 2 hour flight is equivalent to the radiation from a chest x-ray	☐ True ☐ False
Malaria chemoprophylaxis is contraindicated in pregnancy	☐ True ☐ False
Yellow fever vaccine is safe in pregnancy	☐ True ☐ False
Pregnancy is caused by flying more than a mile high	☐ True ☐ False
Giant cell arteritis: Acute blindness occurs in 2% of patients with temporal arteritis without prompt treatment	☐ True ☐ False
Abnormal palpation of the temporal arteries is a predictive sign	☐ True ☐ False
Prednisolone 40mg OD is an adequate dose for all patients with temporal arteritis	☐ True ☐ False
Patients put on a reducing dose of prednisolone can never remember where they are in the reducing process, and neither can you	☐ True ☐ False
Vertebroplasty for vertebral fracture: May improve symptoms but there is no conclusive evidence	☐ True ☐ False
A vegetable pasty is spelt almost the same, and may be just as beneficial	☐ True ☐ False
Nausea and vomiting in pregnancy: Pregnancies complicated by nausea and vomiting are more likely to result in miscarriage	☐ True ☐ False
Ginger and vitamin B6 may help	☐ True ☐ False
Metoclopramide and prochlorperazine are contraindicated in pregnancy	☐ True ☐ False
If you're pregnant and you've got nausea and vomiting and you're flying to India, for God's sake get a seat next to the aisle	☐ Right on ☐ I prefer

the window

Cognitive assessment of older people

BMJ 2-11;343;d5042

John Young, head, David Meagher, professor of psychiatry, consultant psychiatrist, Alasdair MacLullich, professor of geriatric medicine

- Dementia affects 20% of people 80 yo
- Patients with dementia are at greater risk of delirium
- Delirium may result in persisting cognitive deficits
- Every 48 hours spent in delirium is associated with an 11% increase in mortality
- Assessment is based on observation of the patient, a colateral account and standardised tests
- Opioids, benzodiazepines, and dihydropyridines (eg, amlodipine), and possibly antihistamines increase the risk of delirium
- Blood tests for cognitive impairment include: FBC , renal blood, LFTs, tsh, calcium, glucose, CRP, B12, folate.
- Colateral account can been done using AD8 questionnaire
- Untestability is an important sign of delirium
- A score of 7 or less on the abbreviated mental test score suggests cognitive impairment
- PHQ2 has a sensitivity of 100% and a specificity of 77% - a score > 3 suggests depression
- GPCOG is a web-based tool
- If you ask a parent whether he or she has left you anything in his or her will, and the reply is "No", it's proof of dementia

Depression in older adults

BMJ 2011;343; d5219

Joanne Rodda, clinical training fellow in old age psychiatry, Zuzana Walker, reader in old age psychiatry, Janet Carter, senior lecturer in old age psychiatry

- Depression in the over-65s is associated with disability, increased mortality and poorer outcomes in physical illness
- More people aged over 65 commit suicide than any other age group and most have major depression
- Most episodes of depression in the elderly will be a recurrence rather than a first episode
- There is an increased female to male ratio as in younger adults
- Dementia, Parkinson's disease, stroke, diabetes, and CHD increase the prevalence of depression
- Diagnosis is based on clinical interview and collateral history
- Rating scales may overestimate the severity of depression in the elderly
- Somatisation , hypochondriasis, psychomotor retardation or agitation and pychosis are more common in elderly depressed patients than in younger patients
- Depression in later life is associated with cognitive impairment and may increase the risk of dementia twofold
- Vascular depression or depression executive dysfunction syndrome is proposed to cause a disruption of pre-frontal circuitry leading to executive dysfunction and mood disorder
- Structured exercise, CBT, problem solving, and bibliotherapy all lead to improvement in depressive symptoms
- SSRIs are first line treatment
- Response to treatment with antidepressants is delayed in older adults
- When treating depression in Alzheimer's, sertraline and mirtazepine are no better than placebo

- SSRIs may cause drug interactions mediated by cytochrome p450 enzymes , and citalopram, escitalopram and sertraline are safest in this regard
- Hyponatraemia and GI bleeding are side effects of SSRIs in the elderly
- Antidepressants should be continued for at least 6 months for a first episode and at least 2 years if there is thought to be risk of relapse
- Lithium augmentation may be beneficial – confirmed in 2 studies
- ECT is safe and effective
- Psychological therapy is as effective in the elderly as in younger patients
- There is a higher risk of relapse of depression in the elderly
- Two thirds of elderly patients take 3 years to recover from a depressive episode
- In depressed patients >70 yo , there is an increased mortality compared to patients who are not depressed
- You're probably depressed yourself after reading this – I certainly am

Interpreting asymptomatic bacteriuria

BMJ 2011;343;d4780

Martin Cormican, professor of bacteriology, Andrew W Murphy, professor of general practice, Akke Vellinga, epidemiologist

- Dysuria and frequency without vaginal discharge or irritation confirms a uti (probability 90%)
- Treatment of asymptomatic bacteriuria in pregnant women reduces the risk of pyelonephritis later in pregnancy
- Asymptomatic bacteriuria may be present in 20% of healthy women over 80 years old.
- Treatment of asymptomatic bacteriuria is not recommended in non-pregnant women as it does not improve clinical outcome
- Bacteriuria is almost universal in people with long term urinary catheters – only treat if unwell (despite what the Community Nurse says)
- Urine for culture does not have to be mid stream
- A positive urine test for nitrite and either leucocytes or red cells has a positive predictive value of 97%

The challenge of managing coexistent type 2 diabetes and obesity

BMJ 2011;342;d1996

Clifford J Bailey, professor

- Type 2 diabetes is usually caused by insulin resistance, defective insulin secretion, hyperglucagonaemia and coupled with obesity, doubles the risk of death
- A 5 – 10% weight loss can reduce HbA1c by 5 – 11 mmol/ml and increase life expectancy by 2-4 years
- Weight loss improves insulin sensitivity which makes it harder to lose weight
- Orlistat may result in 2-3kg weight loss when compared to placebo
- Bariatric surgery reinstates near normal glycaemia in 50-80 % of patients for several years
- Metformin is weight neutral
- GLP-1 agonists reduced weight by 2-4 kg in trials
- DPP4 inhibitors are weight neutral
- Acarbose helps control weight
- GLP-1 agonists may cause injection site reaction , resistance may develop and possibly increased risk of pancreatitis. Patients must have adequate renal function

Diagnosis and management of ectopic pregnancy

BMJ 2011;342; d3397

Davor Jurkovic, consultant gynaecologist, Helen Wilkinson, director

- The earliest symptom is a brown vaginal discharge soon after a missed period
- Up to 20% of women have no pv bleeding and 10% have no abdominal pain
- Patients with a ruptured fallopian tube and severe intra-abdominal bleeding may present with diarrhoea and vomiting
- Diagnosis of ectopic pregnancy should be considered in all women of reproductive age with sudden onset of severe gastrointestinal symptoms
- Physical examination is of limited diagnostic value
- Methotrexate may be used to manage women with a small ectopic, low HCG levels and few symptoms
- Following salpingectomy intra-uterine pregnancy rates are 38 – 66%, or 62 – 89% after tubal conservation.

Fall assessment in older people

BMJ 2011;343:d5153

Jacqueline C T Close, consultant geriatrician, Stephen R Lord, senior principal research fellow

- Falls account for 4% of hospital admissions
- GPs should ask elderly patients (or their carers) about falls and undertake a mobility screen – such as "timed up and go test"
- If timed up and go test takes longer than 12 seconds, there is an increased risk of falls
- A history of 2 or more falls in the last year should trigger detailed assessment which should include – balance and mobility, vision, syncope, dizziness, postural hypotension, foot pain and deformity, drug review, home environment (OT), cognition. Vitamin B12 and vitamin D levels should be checked.
- Vitamin D supplementation is a simple evidence based intervention in older people in residential care particularly if vitamin D levels are < 50nmol/l.
- There is no evidence of effectiveness of falls intervention in patients with dementia, Parkinson's disease, depression or previous stroke
- Some very grumpy and curmudgeonly GPs seem to think that referring patients to the falls team is a big fat waste of time - but naturally I'm not one of them

Chronic chilblains

BMJ 2011;342; d2708

I H Souwer, general practitioner, A L M Lagro-Janssen, general practitioner

- Chilblains may occur on the fingers, feet, ears or thighs – ulceration may be present
- There may be a link with SLE
- Usually occur when the temperature is below 10 degrees C
- There may be a family history
- Symptoms settle in the spring and if they continue into the summer, they may be chilblain lupus erythematosus
- Raynauds syndrome is not linked
- There is some evidence that nifedipine 60mg a day is better than placebo
- To my astonishment, the well-known fact that chilblains get better if you wee on them was mentioned nowhere in this article

Assessing and helping carers of older people

BMJ 2011;343;d5202

**I D Cameron, head, C Aggar, PhD student, A L Robinson,
professor of primary care and ageing, S E Kurrle, head**

- Caring can lead to a decline in the carer's mental and physical health, especially if caring for those with dementia
- Carers may consider their input as integral to their family duties and decline support
- Asking carers "overall how burdened do you feel" is a useful way to assess carers' distress
- Family support groups have been shown to be effective mainly in the context of mental illness and dementia
- A systematic review shows that respite care had some small effects
- Ongoing contact and support is needed to encourage carers to accept help
- It may be helpful to involve other family members in discussions about support. (On the other hand, families being what they are, it might cause a huge argument.)

Diarrhoea after broad spectrum antimicrobials

BMJ 2011;342;d3798

Chris Settle, consultant microbiologist , Kevin G Kerr, director of infection prevention and control, honorary clinical professor of microbiology

- The number of c difficile cases has dropped substantially in the past 2 years
- The annual incidence in the community is 0.29 per 1000 population
- Quinolones and cephalosporins are most likely to cause c. difficile
- The HPA suggests all diarrhoea specimens in patients over 65 years old should be tested for c.difficile
- Diarrhoea can occur after a single dose of an antibiotic
- C difficile infection may progress to colitis, toxic megacolon and death
- C difficile toxin is usually tested for by enzyme immune assays which give rapid results but may yield false positives and false negatives
- Further investigations should be discussed with the microbiologist if stools are negative but c difficile infection is strongly suspected
- Patients may be treated empirically while awaiting results
- Symptom resolution is sufficient to indicate therapeutic success
- Relapses occur in as many as 20% of cases
- Treatment is with vancomycin 125 mg qds for 10 – 14 days or metronidazole 400mg tds for 10 – 14 days

β blockers for heart failure with reduced ejection fraction

BMJ 2011;343;d5603

Derek G Waller, consultant cardiovascular physician, James R Waller, specialty trainee year 4 cardiology

- Heart failure is characterised by neuro-hormonal stimulation of the sympathetic nervous system and the rennin-angiotensin-aldosterone system
- Beta blockers reduce morbidity and mortality in patients with heart failure due to reduced left ventricular ejection fraction
- Bisoprolol, carvedilol and nebivolol are all licensed for use in heart failure
- Beta blockers reduce morbidity and hospital admissions for heart failure by 34% in addition to the benefits of ace inhibitors
- If LVEF is preserved, there is not enough evidence to recommend beta blockers
- In diabetes, beta blockers may lead to deterioration in blood glucose control
- The mean age in clinical trials of beta blockers in heart failure is 63 years old but the average age that heart failure develops in the community is 76 years old
- Cardioselective Beta blockers do not lead to deterioration in lung function in the short term in mild to moderate asthma
- Beta blockers should be avoided in patients with critical limb ischaemia
- Blood pressure may rise because of improved cardiac function
- Every reduction of 5 beats per minute in people with heart failure treated with beta blockers leads to an 18% reduction in mortality (so if you could lower their heartbeat by 30 per minute they'd presumably be immortal)
- Beta blockers are contraindicated in 2nd and 3rd degree heart block

- Carvedilol or nebivolol may be better in patients with Raynaud's syndrome as they have a vasodilating activity
- The initial dose of beta blocker should be low with gradual titration no more frequently than every two weeks
- ACE inhibitors and beta blockers reduce mortality with a NNT of 26 over one year to prevent one death
- Spironolactone reduces mortality in severe heart failure with a NNT of 9 over 2 years to prevent one death

Effect of β blockers in treatment of chronic obstructive pulmonary disease: a retrospective cohort study

BMJ 2011;342;D2655

Philip M Short, clinical research fellow respiratory medicine, Samuel I W Lipworth, medical student, Douglas H J Elder, clinical research fellow cardiovascular medicine, Stuart Schembri, consultant respiratory physician, Brian J Lipworth, professor of respiratory medicine

- Beta blockers reduce mortality in COPD
- They also reduce respiratory events and morbidity
- Despite initial increase in bronchospasm, chronic escalating doses of beta blockers in patients with asthma reduced airway responsiveness
- Beta blockers should not be withheld for cardiovascular indications, but using them solely for COPD is premature
- Patients with COPD given a cardioselective beta blocker should be observed during administration of the first dose
- Anticholinergic agents should be used if bronchospasm occurs

QUIZ NO 2

Cognitive assessment in the elderly: Delirium may cause persisting cognitive defects	☐ True ☐ False
Every 48 hours spent in delirium increases mortality	☐ True ☐ False
Amlodipine increases the risk of delirium	☐ True ☐ False
Depression in the elderly: Diagnosis is based on assessment tools	☐ True ☐ False
Depressed, demented patients respond well to sertraline and mirtazepine	☐ True ☐ False
Elderly patients respond rapidly to treatment for a depressive episode	☐ True ☐ False
Elderly patients with depression have an increased mortality compared to those without depression	☐ True ☐ False
Asymptomatic bacteriuria: Dysuria and frequency without vaginal discharge or irritation confirms a UTI with a probability of 90%	☐ True ☐ False
Less than 5% of women over 80 years old have asymptomatic bacteriuria	☐ True ☐ False
Asymptomatic bacteriuria in non-pregnant women should be treated with antibiotics	☐ True ☐ False
Asymptomatic bacteriuria is hard enough to say, but typing it is a nightmare	☐ True ☐ False
Type 2 diabetes and obesity: Type 2 diabetes is associated with insulin resistance and obesity	☐ True ☐ False
Weight loss improves insulin sensitivity which makes it easier to lose weight	☐ True ☐ False
Bariatric surgery has little effect on glycaemic control	☐ True ☐ False
Acarbose is basically useless	☐ True

	☐	False
Ectopic pregnancy: Diarrhoea and vomiting may indicate a ruptured ectopic pregnancy	☐	True
	☐	False
Methotrexate may be used to manage a small ectopic pregnancy	☐	True
	☐	False
Following salpingectomy, intra-uterine pregnancy rates are below 50%	☐	True
	☐	False
Acarbose is no use in ectopic pregnancy either. Stupid acarbose. What's the point of it?	☐	True
	☐	False
Falls assessment in the elderly: If a timed "up and go test" takes longer than 12 seconds, the patient has an increased risk of falls	☐	True
	☐	False
If it takes longer than 12 minutes, the patient has gone to sleep	☐	True
	☐	False
There is no evidence that correction of Vitamin D deficiency is worthwhile in elderly people in residential care	☐	True
	☐	False
Falls intervention in patients with dementia, Parkinson's disease, depression or previous stroke is about as effective as a dollop of acarabose	☐	True
	☐	False
Chilblains: Chilblains never occur on the thighs	☐	True
	☐	False
Chilblains may be familial	☐	True
	☐	False
Nifedipine is ineffective against chilblains	☐	True
	☐	False
Chilblains in the summer may suggest a link with SLE	☐	True
	☐	False
Chilblains in your brains are called chilbrains	☐	True
	☐	False
Carers of older people: Care is unlikely to affect the health of the carers	☐	True
	☐	False
Respite care may result in some advantages to carers	☐	True
	☐	False

Carers readily accept support	☐ True ☐ False
Diarrhoea after broad spectrum antimicrobials: The number of cases of C Difficile is increasing	☐ True ☐ False
The annual incidence is 1.5 per 1000 population	☐ True ☐ False
Co-Amoxiclav is the drug most likely to cause C Difficile	☐ True ☐ False
If stool testing is negative, no further action is necessary in spite of clinical suspicion of C Difficile	☐ True ☐ False
Answering questions about C Difficile is V Difficile	☐ True ☐ False
Beta blockers in heart failure: Beta blockers only reduce morbidity and mortality in patients whose heart failure is due to reduced left ventricular ejection fraction	☐ True ☐ False
Beta blockers may lead to poorer blood glucose control in diabetics	☐ True ☐ False
Beta blockers are likely to reduce blood pressure because of improved cardiac function	☐ True ☐ False
Most clinical trials of beta blockers have been done on younger patients than those commonly developing heart failure in the community	☐ True ☐ False
Beta blockers in COPD: Beta blockers increase mortality in patients with COPD	☐ True ☐ False
It is safe to prescribe the first dose of a cardioselective beta blocker (when indicated) in primary care without any special observation	☐ True ☐ False

Antihypertensives in octogenarians

BMJ 2012;344:d7293

Giuseppe Mancia, head, Clinica Medica

- The hypertension in the very elderly trial (HYVET) showed that drug treatment of hypertension in patients 80 years old or older was 'associated with a significant and marked reduction in the incidence of stroke and heart failure (and all cause mortality)'
- A treatment extension of those initially allocated to placebo was studied and found that all cause mortality and cardiovascular disease, mortality was greater in the group who received treatment at the earlier phase.
- The results 'should be interpreted with caution' as the overall and cause specific events were small with large confidence limits of the hazard ratios and limited statistical power.
- Trials in the very elderly present special difficulties such as the 'necessary shorter duration of the observation period' and the confounding role of intercurrent diseases.
- Good BP control is important to achieve reduction in cardiovascular events. However, adherence to anti-hypertensive treatment is low in real life and may be a greater problem in very old age.

Treating sciatica in the face of poor evidence

BMJ 2012;344:e487

Roger Chou, associate professor of medicine

- The most common cause of lumbosacral radiculopathy is disc herniation, which occurs in about 3% of patients with acute low back pain.
- There is little evidence to support the use of any drug for this condition.
- Systemic steroids showed positive effects on short-term pain compared with placebo although the benefit was small.
- It "is necessary to use" 'indirect' evidence by extrapolating from findings of trials evaluating drugs for other conditions and making assumptions about "generalizability" (is this non-evidence-based medicine?)
- Pregabalin, serotonin-noradrenaline reuptake inhibitors and tricylic anti-depressants should be considered.
- Opioids should be reserved for 'severe or intractable' pain and benzodiazepines are not recommended as they "have not been well studied" (does lack of evidence mean lack of effect?)

Benefits and harms of mammography screening

BMJ 2012;344:d8279

Allan Hackshaw, professor

- Screening 'probably does reduce mortality' but the size of the effect is unclear.
- Harms include overtreatment and anxiety caused by false positive results.
- 'Over diagnosis' is the 'proportion of women diagnosed with breast cancer through screening who would not have been detected in the absence of screening, and they therefore receive treatment unnecessarily'
- Benefits from screening appear several years after starting screening.
- Net harms indicated by negative QALYS up to 8 years after screening. "However, these effects are smaller than the net benefits after 10-12 years"
- There is a less than 10% risk of 'over diagnosis"
- The evidence on cervical screening and mortality comes from observational studies only.

Managing motion sickness

BMJ 2011;343:d7430

Louisa Murdin, academic clinical fellow in audiovestibular medicine, John Golding, professor of applied psychology, Adolfo Bronstein, professor of neuro-otology

- Hyoscine is effective as patches, oral preparations and possibly nasal spray.
- There is less evidence for other drugs such as promethazine, cyclizine, budizine or cinnarizine.
- One small trial suggested that ginger was better than placebo in treating motion sickness.
- Mal de debarquement is a persistent imbalance or rocking sensation usually following a sea voyage, especially evident in those who have spent most of the voyage in the bar.

How clinical and research failures lead to suboptimal prescribing: the example of chronic gout

BMJ 2011;343:d7459

Wendy Lipworth, postdoctoral fellow, Ian Kerridge, director, Jonathan Brett, clinical pharmacology and toxicology registrar, Richard Day, professor of clinical pharmacology

- Many patients are prescribed sub-therapeutic doses of allopurinol.
- Probenacid is a safe, cheap alternative to allopurinol in patients who have had kidney stones or serious renal impairment. Probenacid dosage should be gradually increased and patients should remain well hydrated.
- Febuxostat is an expensive xanthine oxidose inhibitor, which may have benefits in a small sub-set of patients. Its long-term safety is uncertain.
- Colchine is used mainly to treat acute gout but can be used in chronic gout to prevent flares occurring when urate-lowering therapy is started.
- Flares can also occur when listening to Led Zeppelin albums.

Malaria

BMJ 2011;342:d1149

Deen M Mirza, assistant professor, Muhammad Jawad Hashim, assistant professor, Aziz Sheikh, professor of primary care research and development

- People who have survived childhood in Malaria endemic countries have usually developed immunity but this declines in the absence of regular exposure.
- P. falciparum in pregnant women can cause prematurity and increases maternal morbidity and mortality.
- Children are at risk of hypoglycaemia and cerebral malaria.
- Insect repellants (DEET) can be used in pregnancy and on children older than 2 months. A permethrim-impregnated mosquito net should be tucked under the mattress and re-impregnated every six months.
- Doxycycline is contra-indicated in pregnancy and breastfeeding and children under 12 years old because of teeth and bone staining.
- Malarone is contra-indicated in pregnancy but can be used in children.
- Mefloquine can be used in children and in the last two months of pregnancy.
- Malarone is only licensed for use for one month in a malarious area whereas other drugs can be used for longer.
- Tell people to seek medical attention if they develop a fever from one week after arrival in a malarious area to twelve months after their return.
- The phrase "malarious area" was invented to prevent people from having to say "malaria area".

Antipsychotic prescribing in nursing homes

BMJ 2012;344:e1093

Jenny McCleery, consultant psychiatrist, Robin Fox, general practitioner

- In 2008 it was suggested that risk of death in patients prescribed anti-psychotics for dementia was greater for typical rather than atypical anti-psychotics
- Information on risk 'must be weighed against the potential benefits of a drug'
- There is no high quality evidence that atypical anti-psychotics help with 'neuropsychiatric symptoms of dementia'
- Anti-psychotics in dementia increase the risk of death and have adverse effects and limited efficacy.
- Guidelines agree that first line treatment of behavioral/psychological symptoms in dementia should be non-drug based. Anti-psychotics should be used for 'severe distress or serious risk to others'
- Despite guidance, anti-psychotics are still widely prescribed.
- In 2008-2009, 18% of patients in care homes were prescribed anti-psychotic drugs. In the community 10.1% of patients with dementia were prescribed anti-psychotics compared with 30.2% in care homes.
- GPs and Psychiatrists may feel 'pressured to prescribe, believing that non-drug approaches are unfeasible' and perceive a failure at social level to provide the environment and resources needed for high quality innovative care.
- "It is fair to say that many doctors think the evidence based guidelines are not adequate for the day to day reality of practice"
- (Perhaps) future research should focus on non-drug treatments and appropriate 'service structures'

Childhood cough

2012;344:e1177

Malcolm Brodlie, academic clinical lecturer in paediatrics, Chris Graham, general practitioner registrar, Michael C McKean, consultant respiratory paediatrician

- 'Parents' reports of the frequency, duration or intensity of coughing correlate poorly with objective observation'
- Chronic cough is defined as 'variably defined' as lasting from 3 – 12 weeks.
- 24% of preschool children have had symptoms 2 weeks after the onset of an URTI.
- Suggestive features of foreign body aspiration include sudden onset of cough or breathlessness.
- Examination should include measure of respiratory rate, heart rate, O2 saturations and temperature.
- Parental concern and clinician's instinct that something is wrong remain important red flags for serious illness (in primary care)
- However, excessive parental concern about a little tickly cough can also be an important red flag for the GP suddenly feeling an uncontrollable urge to punch the parent on the nose
- Refer if
  - acute cough is progressive and severe beyond two to three weeks (with chest signs on examination of LRTI);
  - haemoptesis;
  - or suspicion of
  - cancer,
  - TB or
  - inhaled foreign body.
- Cochrane review found no evidence of effectiveness for OTC (over the counter) cough remedies.
- Apparently 'young children have died of overdoses of OTC drugs for coughs'.
- If a child has pertussis, a macrolide may reduce infectivity.

- Pollen season (February-September roughly) cough may be treated with anti-histamines (or intranasal steroids)
- BTS guidelines define chronic cough as one that lasts over eight weeks.
- Chronic cough in children may be stratified as:
  - normal children
  - children with a clear cause of cough
  - 'Non-specific isolated cough' (dry cough, persistent, not short of breath, no other 'disease' PLUS a normal X-ray)
- Number three may be habit or psychogenic cough.
- In 2007 screening was introduced for Cystic Fibrosis in the UK. This will NOT detect every child with Cystic Fibrosis.
- If there is strong family history of atopy and wheeze on examination suggesting asthma, consider a trail of inhaled steroid and a 'peak flow diary and a defined outcome diary'
- BTS guidelines advise a reversibility test response to a bronchodilator.
- Asthma is unusual under 2 years of age.
- The clinical diagnosis of asthma in children is "difficult" and specialist referral is appropriate if there is uncertainty, or symptoms are difficult to control.
- 'Gastro-oesophogeal reflux is common in infancy and is only sometimes associated with non-specific cough'
- Psychogenic cough is 'honking, bizarre, disruptive and the child is well'. It is worse at night and more prominent in the presence of carers, GPs and homework. It may subside abruptly in the presence of cough-resenting alligators.

Newer antidepressants for the treatment of depression in adults

BMJ 2012; 344:d8300

Simon Hatcher, associate professor of psychiatry, Bruce Arroll, professor of general practice

- Meta-analysis of 35 randomised controlled trials, found that antidepressants were only effective in patients with severe depression
- NNT in severe depression is 4
- For mild to moderate depression treatment depends on patient preference and local resources.
- 'All newer anti-depressants are equally effective' (NICE) although meta-analysis suggests a 'slightly' better response with escitalopram, mirtazapine, sertraline, and venlafaxine (primary and secondary case studies for ONLY 8 WEEKS)
- Escitalopram and Sertraline were most 'acceptable'
- Reboxetine was considered to be ' ineffective and potentially harmful' (which makes you wonder how it got on the market)
- There is no reliable way to 'predict who will respond to which treatment'
- Possible side effects vary from an increase in suicidal thoughts to death from suicide or cardiac arrhythmias.
- Sexual dysfunction is common with second generation antidepressants; highest incidence with citalopram and paroxetine and lowest with mirtazapine.
- NNT for paroxetine and imipramine vary from 4-16 (4 if PHQ-9 is > 20)
- SSRIs increase falls risk (hazard ratio 1.66) and hyponatraemia (hazard ratio 1.5 (ish))
- Tradazone, mirtazapine, and venlafaxine are associated with an increase in 'all cause mortality' (no figures)

- Discontinuation syndrome is more common in SSRIs with a short half-life (paroxetine and venlafaxine) These should be decreased over four weeks.
- NNH for tricyclics is 17 (median) and for SSRIs is 23 (median)
- 30% of SSRI drug interactions reported are due to interactions with tramadol and oxycodone resulting in serotonin syndrome (more common in older people).
- St Johns Wort may decrease levels of SSRIs due to liver enzyme induction. It may react to cause serotonin syndrome.
- NICE says amitriptyline, imipramin , nortriptyline and sertraline are present in breast milk in relatively low levels.
- Venlafaxine may exacerbate hypertension.
- SSRIs increase risk of GI bleeding by reducing platelet aggregation. NNH is 411
- A 5 point reduction in PHQ-9 from baseline is a sign of adequate treatment and a 2-4 point drop warrants an increase in dose.
- If choosing to change medication, consider a different SSRI and then an anti-depressant from a different pharmacological class.
- Fluoxetine should be stopped 4-7 days before starting a tricyclic. MAOIs should be stopped for 2 weeks before starting another anti-depressant.

Cholinesterase inhibitors and memantine for symptomatic treatment of dementia

BMJ 2012;344:e2986

Joanne Rodda, consultant in old age psychiatry, Janet Carter, senior lecturer in old age psychiatry, consultant in old age psychiatry

- The common types of dementia are Alzheimer's, vascular, mixed, Lewy bodies and frontotemporal dementia.
- 4 drugs are licensed in the UK -
- 3 cholinesterate-inhibitors: donepezil, rivastigimine and galantamine
- 1 NMDA receptor partial antagonist: memantine
- There is no evidence to support the use of these drugs in frontotemporal or mild cognitive impairment.
- In mild to moderate Alzheimer's there is a modest mean benefit of 2.7 points on a 70-point scale after 3-6 months treatment with cholinesterade inhibitors.
- There were significant increases in the ADL (activities of daily living scale) with cholinesterase inhibitor treatment.
- The meaningfulness of small changes in clinical scales is 'subject to considerable controversy'
- Recent studies have considered response to treatment as 'reduced worsening' in pooled randomized controlled trials if reduced worsening comparing Cholinesterase inhibitors with placebo suggested a NNT of 6.
- Donezepil (pooled data) trials 'reported significant benefits with severe Alzheimer's' however use of cholinesterase inhibitors in severe disease 'remains outside current license'
- Memantine appears to be no better than placebo for mild Alzheimer's. For moderate to severe Alzheimer's, pooled data suggests benefit after six months on cognition (2.97 points scale), activities of daily living measured 'clinically' and behaviour.

- The Domino study showed 'no significant benefit of the combination of donezepil and memantine compared to donezepil alone in functional or cognitive measures. However, whoever had the double 6 was at a tremendous advantage'.
- The tremendous advantages of a double 6 are also apparent in games of Monopoly and Ludo.
- Most controlled treatment trials average 6 months and drop out rate was 70%
- In vascular dementia, meta-analysis of anti-depressant drugs showed 'no consistent benefit in terms of global measures'
- The most common underlying pathology in dementia is combined Alzheimer's and vascular dementia. Sub-analysis of studies in this group suggests benefit from treatment with Cholinesterase inhibitors.
- Lewy bodies and Parkinson's disease are overlapping conditions associated with marked cholinergic deficits. Treatment may result in reducing worsening of global measures in mild to moderate Parkinson's dementia with an NNT of 7 (24 week trial; rivastigimine) Similar results have been seen with memantine with Lewy bodies or Parkinson's dementia. At present, only rivastigimine is licensed for Parkinson's disease.
- Cholinesterase inhibitors in Alzheimer's have a NNH (number needed to harm) of 7 and donepezil has the fewest adverse effects. Memantine is associated with fewer side effects, although it is associated with constipation, sleepiness, hypertension and headache. That's nothing. You should see the side effects of the other ones.

QUIZ NO 3

Antihypertensives in Octogenarians: The HYVET showed that drug treatment of hypertension in patients 80 years or older was associated with an increase in all cause mortality, and a tendency to lose your slippers	☐ True ☐ False
It's an excellent idea to amend the design of a study after the study has begun	☐ True ☐ False
The benefits of treating hypertension in octogenarians are supported by powerful and unambiguous statistics	☐ True ☐ False
Luckily octogenarians rarely have co-morbidity, and evidence from the trials was based on long-term follow-up	☐ True ☐ False
Even more luckily, octogenarians are highly compliant with antihypertensive medication	☐ True ☐ False
Octogenaria is a small, peaceful island just off Amnesia, famous for its picturesque fallen arches and slipper-groves	☐ True ☐ False
Treating sciatica: More than 96% of patients with acute low back pain do not have a disc herniation	☐ True ☐ False
Steroids are good for relieving sciatica	☐ True ☐ False
It's okay to treat sciatica with drugs that you might have used for other nerve-pain conditions (such as amitriptyline and pregabalin)	☐ True ☐ False
Benzodiazepines are very good at relieving nerve pain	☐ True ☐ False
Benefits and harms of mammography screening: Screening definitely reduces mortality	☐ True ☐ False
Screening may result in unnecessary treatment	☐ True ☐ False
An "over diagnosis" of less than 10% is acceptable	☐ True

	☐ False
The benefits of cervical screening are clear, as they are based on observational studies which prove a causative link	☐ True ☐ False
Managing motion sickness: Hyoscine is good for motion sickness	☐ True ☐ False
There's some evidence that ginger may help motion sickness	☐ True ☐ False
"Mal de debarquement" is the sensation you get when you disembark from a boat only to discover that you're in France	☐ True ☐ False
Gout: Gout may be prevented by allopurinol	☐ True ☐ False
Probenacid may be used as an alternative to allopurinol	☐ True ☐ False
Malaria: Malaria immunity is life-long	☐ True ☐ False
Insect repellants (DEET) should not be used in children under 2 years old	☐ True ☐ False
Malarone is contraindicated in pregnancy, as is doxycycline	☐ True ☐ False
Antipsychotic prescribing in nursing homes: There is clear evidence that atypical antipsychotics can help with "neuropsychiatric symptoms of dementia"	☐ True ☐ False
Guidelines suggest that non-drug treatments for behavioural/psychological symptoms in dementia should be the first line	☐ True ☐ False
There are plenty of resources available in care homes, nursing homes and the community at large to provide non-drug treatment of behavioural/psychological problems in dementia	☐ True ☐ False
Childhood cough: Abnormal respiratory rate, heart rate, oxygen saturations and/or temperature are more important	☐ True ☐ False

red flags for serious cough-related illness than parental concern or a GP's instinct or intuition	
Suspected TB, cancer or inhaled foreign body should be dealt with in primary care	☐ True ☐ False
Cough medicines are effective and safe	☐ True ☐ False
Newborn screening for cystic fibrosis will detect every case from now on	☐ True ☐ False
A bronchodilator/reversibility test is suitable for children of any age	☐ True ☐ False
Antidepressants: Antidepressants work best in people with mild depression	☐ True ☐ False
All newer antidepressants are equally effective, but some are more equally effective than others	☐ True ☐ False
SSRIs increase the risk of falls and hyponatraemia	☐ True ☐ False
SSRIs may interact with tramadol and oxycodone to cause serotonin syndrome	☐ True ☐ False
Despite the name, antidepressants should never be used on ants	☐ True ☐ False
Symptomatic treatment of dementia: Four drugs are licensed in the UK for symptomatic treatment of dementia	☐ True ☐ False
In Alzheimer's there may be significant increases in ADL with cholinesterase inhibitors	☐ True ☐ False
The Domino study showed no benefits of combining donezepil with memantine - but whoever had the double 6 was at a tremendous advantage	☐ True ☐ False
The Domino study was a study of the records of Fats Domino	☐ True ☐ False
Most trials were too long	☐ True

	☐	False
Rivastigmine may be effective in early Parkinson's "dementia"	☐	True
	☐	False
The only difference between NNT and NNH is that one has a T on the end, and the other has an H	☐	True
	☐	False

Navigating the shoals in hypertension: discovery and guidance

BMJ 2012;344:d8218

Morris J Brown, professor of clinical pharmacology, J Kennedy Cruickshank, professor in diabetes, cardiovascular medicine, and nutrition, Thomas M MacDonald, professor of clinical pharmacology

- The aim for public health is to treat all those who would benefit, including those below current definitions of hypertension (who have a log-linear increase in risk with BP), and young patients with stage 1 hypertension.
- In outcome trials the benefits are in those with a clinic BP >140/90 and reductions in BP averaging 10/6mmHg
- Concern must be given to medicalising a healthy population.
- Variability in BP 'might reflect stiffness in large arteries'
- PATHWAY & ACCELERATE studies suggest that combining drugs from the outset improves BP control and reduces adverse events
- There is no evidence to support the guideline assertion that ABPM average daytime BP <150 should not be treated.
- There is no support for the assertion that indapamide or chlortalidone should replace bendroflumethiazide as diuretic choice.
- Clinic measured BP has been studied in 500,000 patients in outcome trials which have shown positive outcomes on mortality, stroke, myocardial infarction and failure. By contrast, there are no randomized trials on the basis of ABPM.
- Previous UK guidance suggests that there is a 10/5 mmHg difference between ABPM and clinic BP measurements.
- There is no significant difference between clinic, home, or ABPM in the 'predictive value' on cardiovascular deaths.
- In PAMELA, a daytime mean ambulatory systolic pressure of 135 (+/- 14) corresponded to a mean clinic systolic pressure of 155 (+/- 22).

- It is common to use the same method for monitoring treatment as for making the diagnosis, which would favour home monitoring
- A daytime systolic average threshold of 150mmHg may translate into 5 avoidable deaths a year for every 1000 patients with hypertension.
- 25-30% of patients have adverse effects with treatment.
- The evidence for diuretic treatment is among the best for any drug for any indication.
- Spironolactone, although advised by NICE, is not licensed for treating hypertension in the UK
- Diuretics are better than calcium channel blockers in preventing heart failure.
- Co-amilozide led to a 44% reduction in coronary events in the MMC Older Adults Trial.

NICE hypertension guideline 2011: evidence based evolution (a response from the NICE authors)

BMJ 2012;344:e181

Richard J McManus, professor of primary care research, Mark Caulfield, professor of clinical pharmacology, Bryan Williams, professor of medicine

- 'Out of office' measurements are better than clinic measurements at predicting risk of cardiovascular events.
- There is 'no evidence of increased risk from white coat hypertension over and above normotension' over a standard 5 year screening cycle.
- 'Ambulatory monitoring was cost effective for people over 60' if re-testing took place annually rather than 5 yearly.
- In 2006, calcium antagonists were shown to be the most cost-effective treatment option unless there is evidence of heart failure in which case thiazides should be preffered.
- Chlortalidone and indapamide are preferred to bendrofluethiazide because ... the UK is 'unique' in its use of bendroflumethiazide.
- As fourth line treatment for resistant hypertension the guideline advised low dose spironolactone.

NICE hypertension guideline 2011: a curmudgeonly response from Dr Hairy

- Why not just put every bugger in the country on medication and have done with it?
- The benefits of putting everyone on tablets for statistical reasons must surely outweigh the downside of taking people who feel perfectly well and convincing them that they've got something the matter with them. After all, a sense of personal wellbeing doesn't have any beneficial effect on your health. Oh no.
- The drug companies must be rubbing their hands with glee.

New recreational drugs and the primary care approach to patients who use them

BMJ 2012;344:e288

Adam R Winstock, consultant addiction psychiatrist and honorary senior lecturer, Luke Mitcheson, consultant clinical psychologist

- New drugs include Ketamine GHB and other stimulants
- Use with alcohol is common and increases risk
- A motivational approach and information about harm reduction 'can be delivered in primary care'
- Suggested approach for patients admitting to drug use:
  - What concerns do you have?
  - Where would you like to go with this? (Proceed with care if the response is "Back to your place")
  - Is there anything I can specifically help you with?
- Internet Resources:-
  - Frank (http://www.talktofrank.com/)
  - Erowid (http://www.erowid.org/)
  - For GPs, Substance Misuse Management in General Practice (www.smmgp.org.uk)
  - The Club Drug Clinic (http://clubdrugclinic.cnwl.nhs.uk/)
  - Drugscope (http://www.drugscope.org.uk/) and the Drugscope Helpfinder (http://www.drugscope.org.uk/resources/helpfinder).

New European guidelines on atrial fibrillation

BMJ 2011;342:d897

Ross J Hunter, research fellow and registrar, Richard J Schilling, professor of cardiology

Priorities in managing atrial fibrillation (in order of their effect on prognosis):
- stroke prevention
- rate control
- rhythm control

The CHA2DS2 –VASc score is as follows:

	Score
Congestive heart failure	1
Hypertension	1
Age >75	2
Diabetes	1
Stroke or TIA	2
Vascular disease	1
Age >65	1
Sex (female)	1

- No treatment is recommended for a zero score and oral anticoagulants are recommended over aspirin for a score of 1
- There is consistent evidence that oral anticoagulants are better than aspirin and (probably) have a similar risk of bleeding.
- The RE-LY trial showed dabigatran 110mg and Warfarin had similar effects on stroke prevention and dabigatran had lower risks of bleeding.
- Less data is available for dabigatran than warfarin and 'it is sensible to continue warfarin in patients who are stable on this drug'
- There are similar outcomes with both rate and rhythm control strategies. Catheter ablation is more effective than

antiarrhythmic drugs or cardioversion in longer-term maintenance of sinus rhythm. Rhythm control should be considered where rate control leaves the patient symptomatic.

- Rhythm control in this instance should never be confused with the rhythm method of birth control. We're trying to settle the patient's heartbeat down, not stimulate it.

- 'Lenient' rate control (<110/min) is acceptable in asymptomatic patients without 'tachycardia cardiomyopathy' in which case 'strict' rate control (<80/mn) is 'mandatory'

- For rate control, B blockers are first choice, calcium antagonists second choice with added digoxin if needed. AV node ablation is an alternative rate controlling strategy.

- Flecainide, propafenone and sotalol are useful for rhythm control but have 'limitations' if there is structural heart disease. In patients with structural heart disease, dronedarone may be an option provided there is no NYHA III-IV heart failure. However, dronedarone is 'less effective' in maintaining sinus rhythm.

- Paroxysmal atrrial fibrillation 'can be eliminated long term' by catheter ablation in 80-90% of patients although 40% will need a repeat ablation.

- In patients with persistent atrial fibrillation catheter ablation is successful in 70-80% of patients with about 50% needing repeat procedures and should therefore be limited to patients who are 'refractory to drug treatment'.

Type 1 diabetes in children

BMJ 2011;342:d294

Keya Ali, consultant paediatrician, Anthony Harnden, university lecturer in general practice and general practitioner, Julie A Edge, consultant in paediatric diabetes

- Secondary nocturnal enuresis is the commonest symptom of new presentation of diabetes in children.
- Secondary enuresis is the earliest symptom of diabetes in 89% of children over 4 years of age.
- Think about diabetes in toddlers with constipation, thrush (oral or vulval), weight loss, headache or lethargy, 'or any acute illness'. In other words, think about it all the time.
- 'Positive predictive value' of the above symptoms is 'not known'
- Diagnose on a single capillary blood glucose test>11.1mmol/l
- Refer the above children to secondary care the same day
- Early diagnosis may prevent death (10/year in UK)
- 10% of children with undiagnosed diabetes have constipation 'secondary to dehydration'
- The incidence of diabetes in children is around 26/100,000 per year. Incidence is increasing by 4% a year in UK and Northern Europe.
- DKA prevalence at diagnosis is around 25% of newly diagnosed diabetic children

Varicose veins

BMJ 2012;344:e667

M-C Nogaro, core surgical trainee, D J Pournaras, specialist registrar general surgery, C Prasannan, general practitioner, A Chaudhuri, consultant vascular surgeon

- Ask about weight loss and rectal bleeding as varicose viens may be due to a pelvic or abdominal mass.
- Refer if lipodermatosclerosis, ulceration, thrombophlebitis or bleeding.
- Uncomplicated varicose veins need no treatment as surgery would not prevent long term complications, particularly ulceration.
- Compression bandaging is the mainstay of treatment for venous ulcers
- Varicose vein surgery will not improve ulcer healing but will reduce the risk of recurrence.
- Class 3 stockings should be worn all day.
- Interventions include: open surgery, radiofrequency ablation, endo-venous laser therapy and foam sclerotherapy.
- Recurrence rates after interventions are 13-30% at 5 years. Surgery will not alter skin changes and may not relieve aching.
- Radiofrequency ablation and laser therapy seal the vein with thermal energy. They are done under ultrasound guidance under local anesthetic and result in quicker recovery.

Diagnosis and management of ANCA associated vasculitis

BMJ 2012;344:e26

Annelies Berden, resident in internal medicine, Arda Göçeroðlu, research fellow, David Jayne, consultant in nephrology and vasculitis, Raashid Luqmani, professor of rheumatology, Niels Rasmussen, senior consultant in otolaryngology, Jan Anthonie Bruijn, professor of immunopathology, Ingeborg Bajema, nephropathologist

- Test for ANCA (Antineutrophil cytoplasmic antibodies) in patients with 'destructive' upper airways disease, pulmonary nodules, renal and pulmonary inflammatory disease, rapidly progressive glomerulonephritis, skin vasculitis.
- There may be leucocytosis, thrombocytosis, raised ESR, normochromic, normocryticanaemia and raised creatinine.
- Chest xray may show infiltrates, nodules, or cavitations.
- ANCA assay can be requested in Primary care particularly in patients with myalgia, weight loss, headaches, hearing loss, nasal symptoms, red eyes, proptosis, nerve palsy, dyspnoea, pyodermagangrenosum, haematuria, hypertension, or decreasing renal function. In other words, everyone >50 years.
- Refer any patient with a positive ANCA test to an 'appropriate specialist'
- ANCA positive tests may relate to SLE, TB, inflammatory bowel disease or even drugs. Or Anchor butter.
- Do ANCA testing if there is 'active urinary sediment, skin vasculitis, chest xray changes and mononeuritis multiplex
- ANCA testing is also indicated in 'subglottic stenosis of the trachea' (?) presenting as slowly progressive dyspnoea
- These conditions may be difficult to recognize without specialist tests.
- Bear in mind that nomochronic nomocytic anaemia and bone pain and renal impairment may also be common presenting

symptoms of multiple myeloma. The symptoms of myeloma may be considered 'normal for age'. Reuleaux stacks and raised ESR may be the only indications of paraproteinaemia. Average rate of ESR is 85 in new cases of multiple myeloma.

- CLL, non-Hodgkins lymphoma and plasmacytomas can be associated with monoclonal praproteins. MGUS occurs in 3% of the population > 50 years old.

The diagnosis and management of aortic dissection

BMJ 2012;344:d8290

Sri G Thrumurthy, honorary research fellow, Alan Karthikesalingam, specialist registrar in vascular surgery, Benjamin O Patterson, clinical research fellow, Peter J E Holt, clinical lecturer in vascular surgery and outcomes research, Matt M Thompson, professor of vascular surgery

- White men >40 years with hypertension or >40 years with Marfans syndrome are at greatest risk
- Patients may present with sudden onset sharp chest pain, loss of consciousness or poor perfusion of end organs
- CT aortography is the first line diagnostic investigation. MR angiography is preferred for surveillance
- Proximal (type A) dissection is treated surgically whereas distal (typeB) is treated medically unless complicated
- Systolic BP should be maintained at 100-200 mmHg in patients with a history of dissection

The management of abdominal aortic aneurysms

BMJ 2011;342:d1384

David Metcalfe, honorary research fellow, Peter J E Holt, clinical lecturer in vascular surgery and outcomes research, Matt M Thompson, professor of vascular surgery

- An AAA is a 'permanent dilation of the abdominal aorta greater than 3 cm in diameter'.
- The mortality after rupture exceeds 80% and accounts for 8000 deaths a year.
- Elective surgery has a mortality of 1-5%
- Patients with a positive family history have the highest risk of AAA formation (possibly eightfold increase in those with an affected sibling)
- Women are less likely than men to have AAA on ultrasound screening although 'high risk populations exist'
- Women with an AAA have an increased risk of rupture.
- The most important modifiable risk factor for AAA formation is smoking (smokers' risk = 7: 1 versus never smokers), and continued smoking is associated with more rapid aneurism expansion.
- There is a weak association between hypertension and accelerated aneurism growth and anti–hypertensives have not yielded 'definitive results'. (Ha! In your face, pharmaceuticals industry!)
- The prevalence of AAA in white men is 10 times higher compared to Asian men. (Ha! In your face, white men!)
- Diabetes may protect against AAA growth but increase risk of rupture.
- Presentation varies from no symptoms to signs of rupture and therefore clinical suspicion and aneurysm screening 'are important in diagnosis'
- AAA may cause distal aneurysms.

- There is a high association between femoral and popliteal aneurysms (62%-85% association).
- US is about 100% sensitive and specific. (How can anything be 'about 100%' for anything? It's either 100% or it's not.)
- 'Strong evidence' exists that a population screening of men over 65 years old is beneficial but not for women. (Ha! In your face, women!)
- The risk of developing a new AAA after a single negative screening is 'small'.
- AAA repairs in men doubled after the introduction of screening (BMY 2009 and Cochrane review 2007)
- The National AAA screening programme (NAAASP) has 'initiated screening in the UK in men aged 65 years old and accepts self referrals from older men'.
- Screening has non-attendance rates of 20%-30%
- The natural AAA course is continued expansion - 2-3 mm average annual growth - and the 'risk of rupture is exponentially related to diameter'.
- Patients with aneurysms less than 5.5 cm should be entered into a surveillance programme, while patients with large aneurysms (> 5.5 cm) should be referred to vascular specialists, or used as party balloons.
- Statins may reduce aneurysm growth (from 3.8 mm – 0.74 a year?) This is controversial.
- Low dose aspirin reduces cardiovascular mortality in patients with AAA.
- Patients should be considered for elective surgical repair once the aortic diameter reaches 5.5 cm. Risk of rupture for AAA 3.0-5.5 cm = 0-1.61 per 100 person years and for aneurysms > 5.5cm = 27/100 person years. Women should be considered for elective repair at diameters > 5.2.
- Endograft repair has lower morbidity and mortality at 30 days: 1.7, vs 4.7 for open repair. (Ha! In your face, open repair!)

What is the most effective way to maintain weight loss in adults?

BMJ 2011;343:d8042

Sharon A Simpson, senior research fellow, Christine Shaw, reader in nursing research, Rachel McNamara, senior trial manager

- Weight loss may be achievable but about a third of weight loss is regained in the following year.
- Only two studies of meal replacement went beyond 2 years.
- Exercise alone did not lead to successful weight loss maintenance.
- Very low energy diets (<600 Kcal) can result in hypokalaemia and cardiac arrhythmia.
- Diet combined with exercise may lead to 20% greater sustained weight loss than diet alone.
- There is no difference between low carbohydrate and low fat diets at 12 months in weight loss in meta-analysis of trials.
- Higher levels of physical activity may help maintain 10% weight loss at 24 months – but this means 275 minutes per week
- Behavioural interventions such as problem solving, peer support, professional support, goal setting and daily self-weighing offer 'significant benefit'. Web based interventions also have significant effects. (If a huge spider grabs you and wraps you in sticky threads, you'll lose weight. It worked for Bilbo Baggins.)
- Orlistat confers a 3.1kg weight loss advantage over diet alone at 24 months but there is a 30-50% attrition rate.
- Bariatric surgery results in about 20% weight loss at two years compared to 1.4 to 5.5 weight loss with non surgical methods.
- Bariatric surgery may be associated with 'adverse effects' such as death (but this does guarantee weight loss). It is recommended as first line treatment for those with a BMI> 50 (that's bariatric surgery, not death).

Vitamin D: some perspective please

BMJ 2012;345:e4695

Nicholas C Harvey, senior lecturer and honorary consultant rheumatologist, Cyrus Cooper, director and professor of rheumatology

- Observational studies suggest that low vitamin D levels are associated with MS, diabetes, colon cancer, CHD, breast cancer, autoimmunity and allergy.
- The UK government advised that pregnant women and children under 5 years old should take 400 iu of vitamin D daily and 46% of health care professionals are aware of this guidance.
- Observational studies cannot prove causality, only association.
- Ricketts and neonatal hypocalcaemic tetany is mostly seen in 'dark skinned' population in the UK. (Ha! In your face, dark skinned population!)
- It is questionable that moderately low vitamin D concentrations in 49% 'of the white population' of the UK is a 'health problem'. What is normal?
- A recent randomized controlled trial suggested that high dose vitamin D supplement may increase the risk of fracture.
- Vitamin D may not be a 'cure all for almost all modern maladies'

Risk assessment of fragility fractures: summary of NICE guidance

BMJ 2012;345:e3698

Silvia Rabar, senior project manager and research fellow, Rosa Lau, research fellow, Norma O'Flynn, clinical director, Lilian Li, health economist, Peter Barry, consultant paediatric intensivist and honorary senior lecturer on behalf of the Guideline Development Group

- The prevalence of osteoporosis in women rises from 2% at 50 years to 25% at 80 years.
- Fracture risk (NICE) should be considered in
  - all women over 65
  - men over 75
  - those who have risk factors including
  - fragility fracture,
  - repeated falls,
  - steroid use,
  - causes of secondary osteoporosis
  - family history of 'hip fracture.'
- Risk assessment includes (NICE)
  - Absolute risk of fracture should be expressed as a percentage (?). FRAX (without a dexa scan) or Qfracture will give you absolute risk estimation.
  - FRAX (www.shef.ac.uk/FRAX) [for people aged 40-90 years]
  - Qfracture (www.qfracture.org/index.php)
- If risk assessment as above suggests intervention then do a DEXA scan and use FRAX again [you will probably/maybe get this calculation in the DEXA report]
- Consider DEXA before sex hormone deprivation for breast or prostate cancer.

- Fracture risk assessment does not include risk associated with anticonvulsants, SSRIs, thiazolidinedione's, PPIs and antiretroviral drugs.

Bisphosphonates in the treatment of osteoporosis

BMJ 2012;344:e3211

Kenneth E Poole, University lecturer, honorary consultant rheumatologist, Juliet E Compston, professor of bone medicine

- Bisphosphonates inhibit bone reabsorption by inducing apoptosis of osteoclasts.
- Large randomized controlled trials in postmenopausal women with osteoporosis have shown significant reduction in vertebral fractures after 3 years of treatment with alendronate (A-ate) risedronate (r-ate), ibandronadte (i-ate) and zolendronate (z-ate) and in non–vertebral fractures with a-ate, r-ate, and z-ate.
- Fracture rate may be reduced in women treated for 3 years with alendronate roughly by half from maximum vertebral rates of 15% and non vertebral rates of 21%
- Trials in men are not designed to look at fracture rates but at changes in bone mineral density. (BMD)
- NNT cannot be calculated reliably between different bisphosphonates
- NICE recommends alendronate as first line treatment for primary and secondary fracture prevention in post menopausal women.
- Risedronate and etidronate are recommended as second line options in women who cannot tolerate alendronate (but this has not been fully explored)
- Musculoskeletal pain, headaches and rashes occur in 1-10% of patients taking bisphosphonates and uveitis may occur in a small number od patients (less than 0.1%)
- A causal relationship between bisphosphonates and jaw osteonecrosis has not been shown.
- Atypical femoral fractures may occur in patients on bisphosphonates prescribed for osteoporosis. They occur with

minimal trauma and heal poorly. The 'estimated incidence' is 1:2000 per year of bisphosphonate use.

- There is no conclusive evidence that bisphosphonates cause arterial fibrillation or oesophageal cancer
- Bisphosphonates should not be prescribed in achalasia or stricture, in patients who cannot stand or sit, in hypocalcaemia, in severe renal impairment or in pregnancy.
- In Barrett's oesophagus, the risk/benefit should be considered 'on an individual patient basis'
- Think about teeth:- does this patient need dental surgery? If so, consider a dental opinion before starting bisphosphonates.
- In secondary prevention in post-menopausal women 'incremental cost-effectiveness ratio (ICER) was less than £30,000 per QUALY. For primary prevention, ICER is £20,000 per QUALY. NICE concluded that at £90 per year for alendronate, it is cost effective in primary and secondary prevention in post-menopausal women and women with previous fragility fracture regardless of BMD.
- Co-prescribe calcium and vitamin D with bisphosphonates unless there is evidence of adequate dietary calcium intake and normal vitamin status.
- Consider measuring BMD every 5 years after treatment. If the BMD remains low or fracture has occurred, continue treatment. In other patients consider a drug holiday. [Consider as evidence based CCG pathway]
- Only denosumab and strontium ranelate reduce non-vertebral fractures as secondary line to bisphosphonates. Raloxifene is recommended by the National Osteoporosis Guideline Group (NOGG) as a second line drug for primary and secondary prevention in postmenopausal women.

QUIZ NO 4

Hypertension guidelines: There is clear evidence that ABPM average daytime BP < 150 systolic should not be treated.	☐ True ☐ False
It's as plain as the nose on your face that indapamide and chlorthalidone are infinitely superior to bendroflumethiazide.	☐ True ☐ False
There are loads of randomised trials of outcome based on ABPM.	☐ True ☐ False
It's even plainer than the nose on your face that ABPM is the method with the best "predictive value" of cardiovascular deaths.	☐ True ☐ False
In PAMELA, a daytime mean ambulatory systolic pressure of 135 (+/- 14) corresponded to a mean clinic pressure of 155 (+/- 22).	☐ True ☐ False
About 3% of patients have side effects with antihypertensives.	☐ True ☐ False
Dr Hairy's curmudgeonly response to this guideline may lead you to suspect that he's more than likely got hypertension himself, but he'd rather not know about it or have it treated.	☐ True ☐ False
New recreational drugs and Primary Care approach: GBH is a commonly used party drug in Maidstone.	☐ True ☐ False
Harm reduction may include advice about use of alcohol in combination with recreational drugs.	☐ True ☐ False
If you're not sure how to help, Frank may be a good resource.	☐ True ☐ False
European guidelines on atrial fibrillation: Basically, if you're a woman with atrial fibrillation you should be on warfarin.	☐ True ☐ False
Rhythm control is much better than rate control in terms of outcomes.	☐ True

	☐	False
Catheter ablation is less effective than antiarrhythmic drugs or cardioversion in maintaining long-term sinus rhythm.	☐	True
	☐	False
Type 1 diabetes in children: Secondary noctural enuresis is the commonest symptom of a new presentation of diabetes in children.	☐	True
	☐	False
It is a good idea to check the capillary blood glucose in any child with constipation, thrush, headache or lethargy.	☐	True
	☐	False
Early diagnosis of type 1 diabetes may prevent death.	☐	True
	☐	False
The incidence of diabetes in children is static in the UK and Northern Europe.	☐	True
	☐	False
Varicose veins: Thanks to NHS reforms, there is no longer any need to refer patients who only have lipodermatosclerosis, ulceration, thrombophlebitis or bleeding related to varicose veins.	☐	True
	☐	False
Compression bandaging is the best treatment for venous ulcers.	☐	True
	☐	False
Surgery for varicose veins will alter skin changes and relieve aching.	☐	True
	☐	False
ANCA associated vasculitis: Consider ANCA assay in patients with multiple symptoms involving multiple systems of the body.	☐	True
	☐	False
Raised ESR, abnormal FBC and raised creatinine may be suggestive of ANCA associated vasculitis.	☐	True
	☐	False
You should refer any patient with a positive ANCA test to an 'appropriate specialist'.	☐	True
	☐	False
You know who the 'appropriate specialist' would be, don't you?	☐	True
	☐	False
Subglottive stenosis presents with slowly progressive dyspnoea.	☐	True
	☐	False

It is easy to distinguish the clinical presentation of ANCA associated vasculitis from that of multiple myeloma. Or scrofula.	☐ True ☐ False
Aortic dissection: If you were to see a 40 year old man with Marfans syndrome, and he said he had a sudden sharp pain in his chest, plus cold peripheries, and while he was saying it he suddenly lost consciousness, it might be a case of aortic dissection.	☐ True ☐ False
The previous question was too long for this sort of quiz.	☐ True ☐ False
Management of abdominal aortic aneurysms (AAA): Elective surgery has a mortality of 1-5%, but after rupture is has a mortality of 80%.	☐ True ☐ False
The prevalence of AAA rupture is greater in Asian men than white men.	☐ True ☐ False
US scan is not a good screening tool for AAA.	☐ True ☐ False
Patients with an AAA > 5.5cm should be referred to a vascular surgeon.	☐ True ☐ False
Maintaining weight loss in adults: Everything worth doing is either illegal, immoral or fattening. And once you get to my age, you can't be bothered with the illegal and immoral stuff.	☐ True ☐ False
Exercise + diet may result in greater weight loss than a deep-fried Mars bar.	☐ True ☐ False
Self-weighing is not effective.	☐ True ☐ False
Non-surgical methods are much better than bariatric surgery in patients with a BMI > 50.	☐ True ☐ False
Vitamin D: The majority of health professionals believe that children under 5 years old should take vitamin D supplements.	☐ True ☐ False
Moderately low vitamin D concentrations occur in 49% of	☐ True

the white population of the UK, and this is a serious health problem.	☐ False
High dose vitamin D supplements may increase the risk of fracture.	☐ True ☐ False
Risk assessment of fragility fractures: You can assess the risk of fracture by using FRAX or Qfracture.	☐ True ☐ False
If FRAX suggests intervention, there is no need to do a DEXA scan.	☐ True ☐ False
Fracture risk should be calculated in all women over 65 and men over 75.	☐ True ☐ False
Bisphosphonates in the treatment of osteoporosis: Bisphosphonates increase the activity of osteoblasts.	☐ True ☐ False
Trials of men treated with bisphosphonates have shown a reduction in fracture rates.	☐ True ☐ False
A causal relationship between bisphosphonates and jaw osteonecrosis has not been shown.	☐ True ☐ False
Bisphosphonates may result in atypical femoral fractures.	☐ True ☐ False
Bisphosphonates should be avoided in patients with severe renal impairment.	☐ True ☐ False
Some patients can have a 'drug holiday' after 5 years of treatment, depending on DEXA scan results.	☐ True ☐ False
A 'drug holiday' doesn't mean sending them off with Cliff Richard on a red bus with a load of hallucinogenics. Apart from anything else, Cliff doesn't like that sort of thing.	☐ True ☐ False

Phimosis in childhood

BMJ 2013;346:f3678

Tamsin Drake, core trainee year 2 doctor, surgery, Jane Rustom, general practitioner, Melissa Davies, consultant urologist

- At age 3 years 10% of boys will have a completely non-retractile foreskin.
- 75% of 5 year old boys have a partially non-retractile foreskin.
- Phimosis may be physiological or due to pathological scarring, which results in a white fibrous ring around the prepuce. In children over 5, the scarring may be labelled as balanitis xerotica obliterans. This label can be attached to the white fibrous ring using a staple or a stitch.
- Ballooning of the foreskin is nothing to worry about and is usually physiological in childhood.
- Balanoposthitis involves the glans and foreskin, whereas balanitis involves the glans only.
- Megaprepuce is rare and congenital and causes a "ballooning scrotal mass". It's a kind of superpower, but you're better off being able to fly or walk through walls.
- If the foreskin is not scarred, reassure parents.
- A 2-8 week course of betnovate 0.05% bd may speed up the natural development process.
- A short course of antifungal cream (or fucidin H) +/- flucloxacillin "may be useful".
- Refer if there is foreskin scarring, more than 3 episodes of balanitis, voiding difficulties or a pinhole meatus.
- The phrase "pinhole meatus" can make even a long-serving GP feel a bit queasy.

Opioids for chronic non-cancer pain

BMJ 2013;346:f2937

Rainer Freynhagen, head of department, associate professor, Gerd Geisslinger, full professor, head of institute, Stephan A Schug, professor and chair

- Data suggests that there is a "continual increase" in the volume of opioids used to manage moderate to severe non-cancer pain. Marketing may have driven the "dramatic increase in use" with more adverse effects including death from overdose.
- Meta-analyses suggest that opioids are good for neuropathic pain, but most guidelines refute this because of the "risk benefit profile".
- The data is even less encouraging for chronic non-neuropathic pain. Generally the benefits are outweighed by adverse events. Opioids "should not be routinely used even for severe osteoarthritic pain".
- There is poor evidence of effectiveness for opioids for osteoarthritis.
- A 2010 Cochrane review of opioids for non-cancer pain (at least 6 months' treatment) suggested pain reduction but a high attrition rate due to lack of effect or side effects.
- There is an increased incidence of falls resulting in fractures in opioid users compared to matched controls – 8% vs 3.2%.
- Respiratory depression is rare but potentially fatal and occurs "most commonly with dosing changes, errors or misuse".
- There are increasing numbers of deaths related to prescription opioids.
- Diversion of prescription opioids can lead to death (diversion occurs in at least 4% of all opioids prescribed in the US).
- Opioids can lead to impaired hypothalamic-pituitary-adrenal axis (dysfunction) leading to OPAID (opioid-induced androgen deficiency) in men.
- Evidence about opioid hyperalgesia is inconsistent.

- Iatrogenic opioid addiction or misuse is about 3.3% of participants in studies, although a more recent study suggested about 1 in 3 patients receiving long term opioid therapy met the "criteria for addiction".
- Cognitive impairment and sedation is possibly minimal with "stable long term doses".
- Use only lower doses in patients with renal impairment.
- The most common adverse effect is constipation – experienced in 40% of patients taking opioids for non-cancer pain.
- NSAIDs and opiates have statistically significant pain reduction but the clinical effect is only minor to moderate.
- Treatment with opioids should be limited to a 4 week trial period. If the treatment is ineffective, deciding not to proceed is a valid option.
- Co-prescribing naloxone in patients on long-acting oxycodone to avoid constipation "is a matter of debate".
- "Most guidelines agree on maximum doses in the range of 100mg of morphine equivalent at which titration should be stopped or requires intense assessment and monitoring".
- Consider the "four As":
  - analgesia
  - activities of daily living
  - adverse effects
  - aberrant drug-taking behaviour
- An exit to long-term therapy is appropriate, especially if aberrant consumption occurs.

The Liverpool care pathway: a cautionary tale

BMJ 2013;347:f4779

Katherine E Sleeman, clinical lecturer in palliative medicine,
Emily Collis, consultant in palliative medicine

- A review of the Liverpool care pathway (LCP) was prompted by press reports of poor end of life care and was led by Julia Neuberger.
- There were many anecdotal accounts from bereaved families of poor care associated with the LCP.
- The panel recommended replacing LCP with "individualised care plans".
- "Without independent prospective evidence from controlled trials, the LCP became unusable. This should serve to warn us of the dangers of the national implementation of tools that are not properly evidence-based."
- Evaluation of "patients' and families' outcomes" including quality of care and death is essential: "analysis of process is not enough".
- End of life care is poorer in hospitals than in hospices.
- Care of the dying is a complex process and requires individual treatment, frequent review and a supportive environment.
- The LCP was intended as a guide but "may have become interpreted by some as a protocol".
- "Healthcare professionals need to be provided with the knowledge, skills and attitudes required to care for dying patients".
- "The ability to communicate uncertainty and share decision making with patients and families is essential."
- The LCP was "suggested as a model of good practice" by the Department of Health, GMC and NICE.

Human to human transmission of H7N9

BMJ 2013;347:f4730

James W Rudge, lecturer, Richard Coker, professor

- H7N9 first emerged in China. The majority of the 133 confirmed cases "seem to be epidemiologically unconnected", with "many" patients reporting recent exposure to live poultry.
- One index case, a 60 year old man, was probably infected at a poultry market. He was nursed for a prolonged period by his 32 year old daughter who also became fatally infected (without exposure to live poultry). There was "almost 100% genetic similarity between the viruses isolated" from father and daughter. The evidence points to transmission from father to daughter.
- There is "no evidence of sustained transmission between humans" – all 43 "close contacts" of the father and daughter tested negative for infection.
- Chinese data suggests that the number of confirmed human cases "is just the tip of the iceberg, as many mild cases are likely to have passed undetected".
- If there are many undetected cases, the mortality is lower than thought, but there may be greater potential for H7N9 to adapt to humans.
- The number of H7N9 cases has fallen abruptly since April 2013 but there may be a resurgence later in the year.

How should we manage fear of falling in older adults living in the community?

BMJ 2013;346:f2933

Steve W Parry, clinical senior lecturer and consultant physician, Tracy Finch, senior lecturer, Vincent Deary, senior lecturer in psychology

- "Fear of falling" includes fear, anxiety, loss of confidence and impaired perception of ability to walk safely without falling.
- It is found in about 50% of older people who have fallen and the same proportion of those who have never fallen.
- Consequences include social isolation, avoidance of activity, increasing frailty and risk of further falls.
- Some evidence supports the use of physical therapy and CBT, although there are no definitive studies.
- A recent study showed that Tai Chi plus CBT showed significant improvement in falls efficacy scale scores compared with Tai Chi alone or fall groups.
- Fear of falling is multifactorial in origin and psychosocial and physiological interventions need to be individualised.
- "Despite the many uncertainties surrounding fear of falling", sustained strength and balance training improves falls risks in general in older adults.

Incretin therapy: should adverse consequences have been anticipated?

BMJ 2013;346:f3617

Edwin Gale, emeritus professor of diabetic medicine

- There may be harmful effects which "have been hidden".
- Incretins act on multiple targets – they are "magic shotguns" rather than "magic bullets".
- Thiazolidinediones modulate numerous genes but may result in weight gain, fluid retention and osteopaenia - and possibly bladder cancer. Two drugs in this class have been withdrawn.
- GLP-1 is a short acting peptide which may act on the heart, kidney, pancreas and thyroid.
- GLP-1 deficiency is (probably) not seen in type 2 diabetes.
- Therapeutic doses are higher (and longer) than physiological doses (and responsiveness).
- Exanatide was the first GLP-1 on the market – from the venom of the Gila-monster. That's a real animal, not something from Star Trek. It's a big lizard that lives in the southwestern United States and northwestern Mexico.
- The Gila-monster goes for weeks without food and conserves energy by gut involution (including the pancreas). There's a lesson for us all there. "Production of exendin-4, a human GLP-1 agonist, causes rapid proliferation of intestinal tissue and a 50% increase in the size of the pancreas when it feeds."
- It was initially hoped that incretins would lead to regeneration of pancreatic ß cells and reverse the progression of diabetes. However, human pancreatic duct cells also carry GLP-1 receptors.
- Pancreatic enlargement has been noted in human studies of GLP-1 agonists and this may be "the mechanism for the occurrence of pancreatitis" which is a class effect of incretins.
- Incretins produce fluctuations in pancreatic enzymes which may be consistent with pancreatic inflammation. Silent pancreatic

inflammation is associated with pancreatic cancer. Incretins may be associated with thyroid tumours.

- It has taken 8 years from the introduction of exanatide for the association with pancreatic abnormalities to come to light.
- "The problem lies in a system that subordinates the public interest to commercial secrecy..."

Restless legs syndrome

BMJ 2012;344:e3056

Guy Leschziner, consultant neurologist, Paul Gringras, professor of sleep medicine and neurodisability

- Restless leg syndrome usually affects (believe it or not) the legs, causing burning, tingling or acheing.
- The sensations are relieved by movement.
- Symptoms are worse at night.
- It may be idiopathic or there may be associations with iron deficiency, pregnancy, uraemia, diabetes, peripheral neuropathy, Parkinsons disease, MS, spinal cord lesions, cardiovascular disease and others – but the strengths (and directions) of these associations "are not clear".
- It is more prevalent in women and older people.
- There are no specific biological markers.
- Sleep disturbance is a common presentation and may delay diagnosis.
- There may be involuntary movements during sleep or while awake. These may be made to alleviate discomfort.
- It may worsen with time, although remission is common.
- Asking "do you have restless legs or troublesome twitches" will identify an (estimated) prevalence of 9-15%.
- Symptoms are "often associated with" ADHD.
- It is the most common movement disorder in pregnancy. Even more common than rushing to the fridge for a peach-flavoured yoghurt.
- There is a higher incidence in Northern Europeans and North Americans.
- Dopaminergic drugs alleviate the symptoms and dopamine antagonists can exacerbate them.
- Serum ferritin levels are inversely related to symptom severity. Iron "seems to have a crucial role in dopamine metabolism".

- Some genes have been identified as possible participants of restless leg syndrome but the findings are inconsistent.
- Management includes iron supplements (if ferritin is below 112pmol/L (50μg/L)), avoidance of SSRIs, antihistamines, metoclopramide and prochlorperazine; also moderate exercise, hot baths, alcohol avoidance and leg massage.
- Drug treatment includes dopamine agonists: ropinirole, pramipexole and rotigotine (the only licensed agents in the UK).
- In meta-analyses, dopamine agents have been shown to reduce symptom severity and improve sleep quality.
- However, dopamine agonists may be associated with the phenomenon of augmentation – earlier onset of symptoms in the day, increased severity of symptoms and spread of symptoms to other parts of the body (arms, trunk or face).
- Restless trunk syndrome could be a particularly nasty problem for the Elephant Man. Or Tom Daley.
- Augmentation is associated with higher doses of dopamine agonists and low ferritin levels.
- Dopamine agonists may increase the risk of impulse control disorders, including gambling, compulsive shopping, hypersexuality and compulsive eating. These disorders may occur in 20% of patients with restless leg syndrome who are taking dopamine agonists.
- If you notice your restless leg patients compulsively buying themselves air tickets, then setting off for Las Vegas with a prostitute and a bag of chips, tell them to lay off the dopamine agonists.
- Rotigotine patches provide therapeutic plasma levels over 24 hours and are advised by the EURLSSG.
- The Movement Disorders Society review of studies before 2007 found evidence that gabapentin is efficacious, and that oxycodone, carbamazepine, sodium valproate and clonidine are "probably efficacious".
- Codeine, tramadol, gabapentin and pregabalin are "recommended as second line treatment" by the EURLSSG taskforce (although unlicensed).
- Clonazepam (off licence) "is suggested as an option for intermittent symptoms that disturb sleep".

- Two recent randomised controlled trials "found pregabalin to be helpful".
- Refer if the following occur: inadequate response, intolerable side-effects, or augmentation.

Hormone replacement therapy

BMJ 2012;344:e763

Martha Hickey, professor of obstetrics and gynaecology, Jane Elliott, senior lecturer, Sonia Louise Davison, senior postdoctoral research fellow

- HRT contains oestradiol, oestradiol 17ß, oestrone, or conjugated equine oestrogen, plus progestogen for women with a uterus.
- Progestogen may be oral, transdermal or via IUD.
- Tibolone has oestrogenic, androgenic and progestogenic actions and can be used for HRT.
- Vasomotor symptoms are the key indication for HRT.
- Vasomotor symptoms "are severe in about 20% of women". The median duration of symptoms is about 4 years.
- HRT may reduce fracture risk, as well as improving vaginal dryness, sexual function, sleep, muscle aches and quality of life.
- Tibolone may be as good or less good at achieving the results above.
- "Starting HRT in women over 60 is generally not recommended".
- In women with premature to early menopause (aged 40-45), current guidelines advise HRT until 50 years old.
- Vaginal symptoms alone do not require systemic HRT and can be managed with local oestrogens.
- There are no large randomised controlled trials of the benefits and harms of HRT in women around the normal age of menopause. The estimated risks come from subgroups of the Women's Health Initiative Study and the Nurses' Health Study.
- A systematic review found that low dose oral or transdermal oestradiol preparations "did not increase the risk of thromboembolism in low risk populations".
- HRT is contraindicated in patients with previous venous thromboembolism (VTE) or in patients at high risk of VTE.

- "HRT increases the risk of stroke" but the risk "may be lower with transdermal HRT at doses of 50μg or less", but this has not been shown in randomised controlled trials.
- Tibolone also increases the risk of stroke.
- HRT and tibolone should be avoided in "women at high risk of stroke".
- In women aged 50-59 years "there was no statistically significant cardiovascular risk or harm".
- The Women's Health Initiative (WHI) study reported an excess breast cancer risk from HRT, which equates to about a 0.1% increase in breast cancer.
- Combined HRT also increased breast density and "the risk of having an abnormal mammogram... Studies are consistent in showing a greater risk of breast cancer with combined HRT than with oestrogen alone."
- Continuous combined HRT may not increase the risk of endometrial cancer (in appropriate dosage), but sequential HRT may increase risk. Tibolone does not increase the risk of endometrial cancer.
- "HRT increases the risk of cholecystitis".
- Avoid HRT in patients with
  - a history of breast cancer (includes tibolone)
  - a history of venous or arterial thromboembolic disease
  - uncontrolled hypertension
- Consider HRT in those at high risk of fracture.
- Perimenopausal women may need contraception. In those without containdications, combined oral contraceptive pills will treat vasomotor symptoms and reduce fracture risk.
- In perimenopausal woman consider cyclic HRT or (in women under 50 years) low dose combined oral contraceptive.
- "In women who are 1-2 years postmenopausal and wish to avoid bleeding, consider continuous combined HRT or tibolone."
- Cessation of HRT leads to recurrent symptoms for up to 50% of women.
- No clear consensus has emerged on how to discontinue HRT, and symptoms may recur if HRT is stopped slowly or suddenly.

- The jury is out on whether it's possible to suddenly stop HRT slowly, or slowly stop it suddenly.
- Slowly advising women to stop suddenly may or may not help.
- Suddenly advising them to stop slowly would probably just give them a surprise.
- Effective non-hormonal preparations for vasomotor symptoms include venlafaxine and SSRIs.
- SSRIs that induce CYP2D6 should be avoided on women with tamoxifen as they may interfere with tamoxifen metabolism (this means paroxetine and fluoxetine – citalopram is better).
- Gabapentin is the only non-hormonal product shown to be as effective as low dose oestrogen for vasomotor symptoms.
- "Clonidine is mildly effective."
- CBT, mindfulness therapy and relaxation therapy may help with vasomotor symptoms.
- Vaginal oestrogens can be used safely in the long term without progestogens.

Are accident and emergency attendances increasing?

BMJ 2013;346:f3677

John Appleby, chief economist

- Jeremy Hunt has blamed lengthening waiting times on GPs' changes in out of hours commitment in 2003-2004, and on patients' difficulties in "obtaining speedy appointments with their GPs".

- Conservative MP Chris Skidmore (Skidmore Underpants) has suggested that immigrants were to blame, although there is evidence that immigrants are "lower than expected" users of secondary care.

- The "obvious reason for recent problems... is that demand is rising".

- In 2003-04 emergency department attendances included "major" units only. Smaller units including walk-in centres and minor injury units were introduced around this time and were then included in attendance figures.

- Much of the increase in 2003-04 due to "previously unrecorded attendances at minor injury units and walk in centres".

- Attendances at "type 1" (major units) "have remained more or less unchanged".

- "Over the last 30 months" the increase in A&E workload has started to level off. The trend increase between November 2010 and May 2013 works out to about 1.9% - "not huge".

- Part of the problem is that A&E departments are "near to capacity" and this is compounded by bed pressure which delays transfer and "further treatment".

- There has been a fall in A&E waiting times in the first few weeks of May 2013, partly due to seasonal factrors and "a greater focus by hospitals and NHS England on dealing with the problem as waiting times escalated".

- Is the problem solved? Watch out for "NHS England's review of the urgent care system – due to report this September".
- The only person who's a bigger git than the current Health Secretary is the one we're going to get next.

Government's plans for universal health checks for people aged 40-75

BMJ 2013;347:f4788

Felicity Goodyear-Smith, academic head

- There are 15 million people aged 40-75 in the UK and the goverment has implemented plans for them to "receive regular, free health checks".
- Since April 2013, local authorities have overall responsibility for providing these checks, including assessment of "alcohol consumption and dementia" in low risk people.
- Patients with a known cardiovascular risk of 20% or more are excluded (including people taking statins).
- About 50% of people accept their invitation for a health check.
- The aim is to "reduce morbidity and preventable deaths".
- "A 2012 Cochrane review concluded that such health checks reduce neither morbidity nor mortality."
- Public Health England says that there is "no time to wait for long term trials to provide definitive supportive evidence".
- The question remains: do health checks meet the criteria for screening? - specifically, "do benefits outweigh harms?"
- Does the health check programme's estimated cost of £332 million per year represent value for money?
- Health checks on low risk people have a high chance of false positives.
- People who come for health checks are "likely to be higher socioeconomic status, with fewer health risks".
- Ticking the health check box may not result in appropriate follow-up and action.
- Patients may choose not to make lifestyle changes if their risk reduction is negligible, or conversely may overestimate the benefits of drugs such as statins.
- There may be a high attrition rate.

- A GP's time "is a finite resource". (That's why we make the Health Care Assistants to do the checks, heh heh!)
- Despite their doubts about the usefulness of health checks, surgeries often do them anyway, because otherwise the local pharmacy would do them, and every time they found anything significant they'd send the patients straight up to the surgery.
- Another reason GPs do the health checks is because they're worth £25 a go.

Smoking cessation strategies

BMJ 2012;344:e1732

Simon Chapman, professor of public health, Melanie Wakefield, director

- Many smokers do not use nicotine replacement therapy (NRT) correctly.
- Professional support may improve outcomes.
- A quitline involves the least inconvenience and cost as an intervention for supporting smoking cessation.
- Only about 6% of smokers call a quitline. Only 7% of those who contact a quitline agree to set a quit date.
- "Online support is increasingly popular."
- "There is probably an upper limit to consumers' acceptance of more intensive support."
- A recent study showed that smokers randomised to the offer of free NRT had slightly lower cessation rates than those participating in standard telephone interaction.
- Studies of real world cessation "mostly show that those who quit smoking unassisted have better long term success rates than those who use medication".
- If quitting unassisted has better outcomes than quitting with assistance, then perhaps quitting in the teeth of discouragement from your GP would be better yet. Maybe the best way to help your patients give up smoking is to tell them they're not allowed to. Since no studies have shown this to work, and there's absolutely no evidence for it, it's surprising that the Government hasn't adopted it as a national policy yet.

QUIZ NO 5

Phimosis in childhood: Only 10% of 5 year old boys will have a partially retractile foreskin.	☐ True ☐ False
Scarring may cause a white fibrotic ring around the prepuce.	☐ True ☐ False
"Ring around the prepuce" is a popular nursery-rhyme dating back to the Middle Ages.	☐ True ☐ False
Balanoposthitis and balanitis are pretty similar.	☐ True ☐ False
Ballooning of the foreskin is nothing to worry about, but a ballooning scrotal mass may warrant a trip to the GP.	☐ True ☐ False
Opioids for non-cancer pain: Meta-analysis suggests opioids are good for neuropathic pain.	☐ True ☐ False
Opioids are good for severe osteoarthritic pain.	☐ True ☐ False
Lack of effect or side effects limit the value of opioids for non-cancer pain.	☐ True ☐ False
Opioids reduce the risk of falls.	☐ True ☐ False
Opioids increase the risk of falls, but you neither know nor care that you've fallen.	☐ True ☐ False
Confusingly, opioids may cause OPIAD.	☐ True ☐ False
Up to 1 in 3 patients on long term opioids meet the criteria for addiction.	☐ True ☐ False
A maximum daily dose of 200mg of morphine equivalent is recommended.	☐ True ☐ False
The Liverpool Care Pathway: The LCP was thought to be	☐ True

a good model by the Deparment of Health, the GMC and NICE.	☐ False
There were anecdotal accounts of poor care with LCP from bereaved families.	☐ True ☐ False
It is not a good idea to implement tools nationally if they are not evidence based.	☐ True ☐ False
End of life care is poorer in hospitals than in hospices (no mention of home care...)	☐ True ☐ False
It is easy for healthcare professionals to be "provided with... the attitudes required to care for dying patients."	☐ True ☐ False
Human to human transmission of H7N9: Most human cases occurred after exposure to poultry.	☐ True ☐ False
Human to human transmission does occur, but the numbers are paltry.	☐ True ☐ False
There may be many mild undetected cases.	☐ True ☐ False
Fear of falling in older adults: Fear of falling increases the risk of falling.	☐ True ☐ False
Paradoxically, you're no more likely to be afraid of falling if you've fallen than if you haven't fallen.	☐ True ☐ False
Numbers of elderly people are falling.	☐ Flue ☐ Tralse
Physical therapy, CBT and Tai Chi may help to reduce falls risk.	☐ True ☐ False
Incretin therapy: Exanatide was the first GLP-1 on the market, and came from the venom of the Gila monster.	☐ True ☐ False
GLP-1 deficiency is seen in type 2 diabetes.	☐ True ☐ False
GLP-1 therapeutic doses are the same as physiological doses.	☐ True ☐ False

Pancreatic enlargement has been noted in human studies of GLP-1 agonists and this protects against pancreatitis.	☐ True ☐ False
There's a Gila monster at the door, and he wants his venom back.	☐ True ☐ False
Restless leg syndrome: Gabapentin is efficacious, but Lily's the Pink's medicinal compound is most efficacious in every case.	☐ True ☐ False
Refer if inadequate response, intolerable side effects or augmentation occur.	☐ True ☐ False
There may be involuntary movements during sleep, sometimes associated with a loud shout of GOAL!	☐ True ☐ False
Serum ferritin levels are inversely related to symptom severity.	☐ True ☐ False
Iron supplements are advised if ferritin < 112 pmol/L.	☐ True ☐ False
Dopamine antagonists are effective treatments.	☐ True ☐ False
Dopamine agonists may worsen symptoms (augmentation).	☐ True ☐ False
Dopamine agonists reduce the risk of gambling. Well, they may do. It's an even chance, but worth a flutter.	☐ True ☐ False
Hormone replacement therapy: Vasomotor symptoms are severe in about 50% of women and have a median duration of 10 years.	☐ True ☐ False
There are loads of large randomised controlled trials of the benefits and harms of HRT on women around the normal menopause.	☐ True ☐ False
HRT and tibolone should be avoided in women at high risk of stroke.	☐ True ☐ False
The WHI study reported an excess breast cancer risk with HRT of about 10%.	☐ True ☐ False

Cessation of HRT always leads to recurrent symptoms in all women.	☐ True ☐ False
Venlafaxine and SSRIs are effective for vasomotor symptoms.	☐ True ☐ False
A&E Attendances: Jeremy Hunt blames GPs for lengthening A&E waiting times	☐ True ☐ False
Chris Skidmore blames immigrants for lengthening A&E waiting times	☐ True ☐ False
Boris Johnson blames the Martians for lengthening A&E waiting times	☐ True ☐ False
The Martians blame Jeremy Hunt for lengthening A&E waiting times	☐ True ☐ False
Attendances at "type 1" (major) units have actually not changed much	☐ True ☐ False
The problem is compounded by bed pressure, which is partly caused by reductions in the number of hospital beds	☐ True ☐ False
Universal health checks: Only high risk patients will be offered health checks	☐ True ☐ False
A 2012 Cochrane review concluded that health checks reduce neither morbidity nor mortality	☐ True ☐ False
Public Health England are right to implement a programme that has no definitive supporting evidence	☐ True ☐ False
Smoking cessation: Only about 6% of smokers call a quitline, but the majority of those who do call will agree a quit date	☐ True ☐ False
There is no upper limit to consumers' acceptance of more intensive support	☐ True ☐ False
Free NRT is likely to improve quit rates when compared to standard quitline interaction	☐ True ☐ False

Epistaxis

BMJ 2012;344:e1097

Omar Mulla, specialty registrar, Simon Prowse, specialist registrar, Tim Sanders, academic general practitioner, Paul Nix, consultant ear, nose, and throat surgeon

- Posterior nose bleeds run into the throat or out of both nostrils, whereas anterior nose bleeds usually run out of one nostril.
- Ask about nose picking (This is particularly good fun if you're talking to a posh old lady.)
- Check BP, especially if the patient is hypertensive. Consider clotting disorders and medication (aspirin, clopidogrel, warfarin) and possible drug interaction.
- Consider clotting screen and FBC.
- "Facial pain or deep otalgia with epistaxis may be the first sign of a nasopharyngeal tumour".
- In a young male consider juvenile nasopharyngeal angiofibroma and ask about nasal obstruction, headache, rhinorrhoea and anosmia. Juvenile nasopharyngeal angiofibromas are rare benign tumours that tend to bleed. They occur in the nasopharynx of prepubertal or adolescent males.
- 95% of nose bleeds arise from Little's area.
- If actively bleeding, ask the patient to lean forward and apply pressure to the soft cartilaginous part of the nose for 10 minutes. If the bleeding does not stop, refer to hospital ENT department, and order a new carpet.
- If the bleeding has stopped, examine the nose with an otoscope to look for a bleeding vessel which looks like a red dot on pale mucosa. Cautery can be attempted using a silver nitrate stick. If there is no facility for cautery, refer to ENT.*
- If the history suggests anterior epistaxis and no vessel is seen, give a 2 week course of naseptin cream or petroleum jelly and ask the patient to return if bleeding persists.

* Note: For "refer to ENT" please read "Attempt to reach ENT on-call Registrar, give up after 15 minutes, send patient to A&E", or "Speak to ENT on-call Registrar, have argument with same, send patient to A&E" or "Send patient straight to A&E on the basis that anything else is too much hassle".

Dietary fats and breast cancer risk

BMJ 2013;347:f4518

Kay-Tee Khaw, professor of clinical gerontology

- A combination of 10-fold international variation in the incidence of breast cancer and changing rates of breast cancer in migrants from low-risk to high-risk countries suggests that environmental modifiable risk factors are involved.
- There is a strong international correlation between national per capita fat consumption and breast cancer rates, but the role of dietary fat is still unclear.
- Zhong et al investigated the association between fish intake and n-3 polyunsaturated fatty acids (n-3 PUFAs) and the risk of breast cancer (meta-analysis).
- There are plausible biological mechanisms for a preventive role for n-3 PUFAs in breast cancer.
- Zheng et al found no significant association between intake of fish or α-linolenic acid and breast cancer. However, the category with highest dietary intake of marine n-3 PUFAs was associated with a 14% reduction in breast cancer risk compared to the category with lowest dietary intake.
- There are confounding factors which cannot be excluded completely. It is also difficult to quantify dietary fat intake and also different types of fat.
- The is an inverse association with marine n-3 PUFAs but not plant-based n-3 PUFAs.
- The association between different fatty acids and breast cancer may depend on food sources. It is hard to distinguish between the effects of nutrients and that of the food itself or dietary patterns.
- Despite the lack of association with fish intake, fatty fish is still the main source of marine n-3 PUFAs.
- The WHI (Women's Health Initiative) study reduced total fat intake by 9% and reported a relative breast cancer risk of 0.91 in the intervention group. The trial had insufficient power to

confirm if this was consistent with any effect of dietary fat on breast cancer risk or not.

- N-3 PUFAs are not to be confused with N3 Poofters, a group of gays living in the Finchley district. There's no evidence that N3 Poofters reduce the risk of breast cancer. Sorry, boys.

Action plans for patients with chronic obstructive pulmonary disease

BMJ 2012;344:e1164

Graeme P Currie, consultant chest and general physician, David Miller, clinical research fellow

- Patients with worsening or more frequent exacerbations of COPD are common and result in increased attendance in primary and secondary care.
- "Such exacerbations will probably increase in the future as populations age [and] rates of worldwide smoking are maintained".
- Patients with repeated exacerbations have "an accelerated decline in lung function and health status, impaired quality of life... and higher mortality".
- Bucknall and colleagues investigated whether "self management in COPD – keeping a symptom diary and use of an action plan" can reduce hospital admissions.
- This was a large RCT of patients with severe COPD (mean FEV1 = 41%) who were admitted to (Scottish) hospitals. Patients received "usual care" or "symptom monitoring intervention".
- The idea was to treat the intervention group with steroid and antibiotics "promptly" during exacerbations.
- "Surprisingly" there were no significant differences between the groups in terms of hospital admissions, death over 12 months, or quality of life parameters.
- Sub-group analysis (not in the study design) deemed some patients to be "successful self-managers".
- A large proportion of those "who did participate, failed to provide adequate questionnaire data".
- Women were less likely to complete symptom diaries, more likely to be current smokers and to be living in deprived communities.

- More than 40% of participants were current smokers – "this highlights the importance of rigorous advice on smoking cessation... [and] a worrying disregard for such advice".
- Conclusions? - "A generic self management approach for all patients with COPD... is unlikely to be effective in terms of resources, cost, or time".
- Curmudgeonly remarks? - Perhaps one day we could try researching things first, to see if they actually work, and implementing them afterwards.

Management of renal colic

BMJ 2012;345:e5499

Matthew Bultitude, consultant urologist, Jonathan Rees, general practitioner with special interest in urology

- The life-time risk of urinary stones is about 12% in men and 6% in women. Bah! And they still expect you to let them have your seat on the bus!
- 80% of stones contain calcium, with uric acid stones comprising 7% of the total (especially common in obese patients). A further 7% are "infection stones".
- Dehydration, sun exposure, diet, anatomical urinary abnormality, primary hyperparathyroidism, renal tubular acidosis, myeloproliferative disorders and gout predispose to renal stones.
- "Renal colic" may have intrinsic or extrinsic causes, eg. lymphadenopathy, blood clots, sloughed papillae; sickle cell disease, diabetes, or long term use of analgesics.
- The classical symptoms include loin pain, referred pain and strangury (urgency, frequency of micturition and straining).
- Nausea and vomiting are common, as is restelessness.
- "A leaking abdominal aortic aneurysm can mimic left sided renal colic". Patients at high vascular risk "need immediate referral for imaging".
- If community management is possible, "urgent imaging must be arranged". Expert consensus suggests that 7 days is "the maximum acceptable interval" – otherwise, hospital admission is necessary.
- Any sign of urinary sepsis "must be excluded, as an obstructed kidney is an emergency" and patients may develop overwhelming sepsis.
- Decision-making regarding investigation of "renal colic" should not be based on the presence or absence of haematuria on dipstick-testing.

- Serum calcium and urate are mandatory tests in patients with proven stone formation (eg. a Druid, or an old man with a rockery in his garden).
- Patients going to A&E with renal colic will probably have immediate imaging, although this is not needed if the pain has settled, there is no sepsis, and the patient has a contra-lateral working kidney.
- Non-contrast CT (NCCT) is the imaging method of choice for "acute flank pain".
- US scan may miss stones between the PUJ and VUJ. However, this is the first line test in pregnancy and children.
- NCCT is often required if plain radiography and US are inconclusive.
- The VUJ is the commonest site of presentation (about 60% of stones in a study of patients admitted to A&E with renal colic).
- NSAIDs are better than opiates in pain relief. The British Association of Urological Surgeons advises diclofenac as first line treatment. Pethidine causes vomiting.
- 2L oral hydration per day is probably a good idea.
- Local warming and acupuncture is probably as good as IM analgesia.
- Sieving the urine to find a stone (peeing into a sock) and biochemical analysis in first time stone formers will avoid the need for further imaging if symptoms have settled.
- Don't wear the sock again afterwards. Not until it's dried out, anyway. Alternatively, pee into somebody else's sock. But not while he's got it on. Unless that's the kind of thing he likes.
- Basically, small stones pass quicker.
- "Several studies" have shown that alpha 1 blockers or calcium antagonists may increase the likelihood of stone passage.
- Buying a house made out of stone may also increase the likelihood of a stone passage.
- Meta-analysis suggests an NNT (Number Needed to Treat) of 4.
- Alpha 1 blockers seem to be much more effective than calcium antagonists. These drugs are "off-label" (unlicensed).
- Most stones cause "partial obstruction", but if a stone is not passed in 4-6 weeks "it is unlikely to do so".
- Surgical intervention is needed in cases of:

- obstructed infected kidney
- obstruction of a solitary kidney
- bilateral renal obstruction
- uncontrolled pain
- Decompression of an infected-obstructive kidney needs percutaneous nephrostomy or retrograde ureteric stent. There is "no significant difference between the two interventions".
- "Patient preference is paramount" for choice of treatment, especially if personal commitments or foreign travel may influence the situation. Travel insurance may be invalidated.
- The main treatments are:
  - extracorporeal shock wave lithotripsy (ESWL) or
  - ureteroscopy
- The above treatments may not be available in all units.
- However, since the success rates are so high, "patients can opt for the less invasive option of ESWL if available".

Familial breast cancer: summary of updated NICE guidance

BMJ 2013;346:f3829

D Gareth Evans, professor of genetic medicine, John Graham, director and consultant in clinical oncology, Susan O'Connell, researcher, Stephanie Arnold, information specialist, Deborah Fitzsimmons, health economist

- Familial breast cancer occurs in people with one or more family members affected by breast, ovarian or primary peritoneal cancer.
- About 5% of all breast cancers are attributed to inherited mutations of high risk genes – BRCA1, BRCA2 and TP53.
- If a patient with no personal history of breast cancer has concerns about relatives with breast cancer, take a first and second degree family history to assess risk.
- Refer if
  - one first degree female relative has developed breast cancer at less than 40 years old
  - one first degree male relative has developed breast cancer at any age
  - one first degree relative has developed bilateral breast cancer where the first primary was diagnosed < 50 years of age
  - two first degree relatives or one first and one second degree relative have developed breast cancer at any age
  - one first degree or second degree relative has developed breast cancer, and one first or second degree relative has developed ovarian cancer, at any age
  - three first or second degree relatives from the same side of the family have developed breast cancer at any age
- In secondary care a carrier probability calculation method will be used.
- Patients with a combined BRCA1/BRCA2 probability > 10% will be referred to specialist genetic clinics.

- Family members of a patient with a personal history of breast or ovarian cancer will be offered genetic testing.
- MRI is more cost effective than mammography for high risk women with a known BRCA or TP53 mutation of > 30% probability of being a mutation carrier.
- Annual MRIs will be offered to women aged 30-49 with/without a personal history of breast cancer if they have BRCA mutation or > 30% risk of being a BRCA carrier.
- Tamoxifen and Raloxifene reduce the risk of breast cancer.
- Tamoxifen for 5 years can be offered by specialists to women at high risk, whether pre-menopausal or post-menopausal.
- In secondary care, discuss the risks and benefits of mastectomy in women with known or suspected BRCA or TP53 mutations.
- Secondary care should also discuss the risks/benefits of bilateral salpingo-oophorectomy in women with BRCA1, BRCA2 or TP53 mutation – reduced risk of breast and ovarian cancer, but negative effects of a surgically induced menopause.
- HRT may be an option for women after oophorectomy if they are 50 years old or less with no personal history of breast cancer.
- The lower thresholds for genetic testing in this guideline will increase the number of referrals to specialist genetic clinics.

Should GPs be fined for rises in avoidable emergency admissions to hospital?

- Bollocks they should

...Sorry, I'll try that again.

Should GPs be fined for rises in avoidable emergency admissions to hospital? Yes

BMJ 2013;346:f1389

Martin McShane, director for domain

- Many admissions are the result of exacerbations of long term conditions and/or "failure of coordinated care" in frail elderly people "with comorbidities needing proactive care from primary, community and social care".
- You can tell what a twit this bloke is, because he keeps using the word "care" more than once in the same sentence.
- We need to "work collectively across the continuum of health and social care" if we are to "reverse the trend in acute admissions".
- "Success will require a range of enablers, levers and incentives to help leaders to change attitudes and behaviours".
- Levers and enablers to help leaders change behaviours? Martin McShane, director for domain? What is this guy, a rap artist?
- "In my experience*, through better use of data, planning, service redesign, contracting and monitoring performance... it will be possible to improve quality while managing costs".

* Experience gained whilst on tour with Snoop Dogg and Slim Shady, no doubt.

Should GPs be fined for rises in avoidable emergency admissions to hospital? No

BMJ 2013;346:f1391

Chaand Nagpaul, general practitioner

- Ah, now here's a sensible chap if ever there was one. What a lovely man! It's a pleasure to read anything by him!
- Reducing emergency admissions is a desireable and worthy aim.
- Emergency admissions represent 65% of hospital bed days in England, at an annual cost of £11 billion.
- The health data company Dr Foster estimated that 29% of these admissions are "potentially avoidable and amenable to interventions in the community".
- Dr Foster also says that overoccupancy of hospital beds is at "breaking point", risking patient safety.
- Emergency hospital admissions have risen by 37% over the last 10 years.
- The NHS is required to save £20 billion by 2015 and avoiding emergency admissions is a key policy to deliver this.
- The recently announced "quality premium" will reward CCGs if they are able to reduce or maintain emergency admission rates within a fiscal year.
- Financial incentives paid to GPs as part of the practice based commissioning scheme (2005-2011) "were unable to stem the rise in emergency admission rates". (We knew it wouldn't work. We thought we might as well take the money, though.)
- There is no evidence that "risk stratification" and "case management" are effective in reducing emergency admissions.
- "The fundamental flaw in linking financial payments to GPs to emergency hospital admissions is that the GP is only one player in a multiplicity of factors" influencing such admissions. "It is therefore inappropriate for GPs... to be held responsible for emergency admission rates."
- One of the multiple factors influencing emergency admission rates is "internal hospital organisation and admission policies".

- 70% of weekly hours "fall outside the control of general practices". (Mind you, the Government seems determined to make GPs responsible for the whole lot.)
- The 111 service will "also refer patients directly to hospitals", bypassing Gps.
- GPs have no control over RTAs.
- Except, of course, if they cause them themselves, by careering around the countryside at 90mph because they've got four home visits and they haven't had lunch yet. But that never happens.
- The four hour maximum wait in A&E "may" have increased "short term admissions".
- Small numbers of patients on GP lists "result in variations in admission rates by chance or volatility... (for example, infection outbreaks)."
- Hospitals have no incentive to reduce admissions "nor to collaborate with Gps".
- 2 recently published, high quality RCTs of interventions designed to keep people out of hospital "showed increased deaths amongst the intervention groups".*
- Basically the purchaser-provider split is exacerbating the problem. "A system of collaboration and shared financial 'ownership', aligning primary, secondary and social care" may address the wider determinants "that influence hospital admissions".
- GP reward/penalisaton "on the basis of emergency admission rates is likely to squander precious public resources on unproved ideology".
- Well done, Dr Nagpaul! First rate! I couldn't have put it better myself!

* Perhaps not surprising, as one of the interventions concerned was to pay a hit-man from the Mafia to rub people out if they got too close to the A&E Department.

The hospital bed: on its way out?

BMJ 2013;346:f1563

John Appleby, chief economist

- Since 1979 NHS beds for acute care have fallen by 35%, for maternity by 58%, for geriatric care by 65% and for mental health by 74%.

- Allowing for population increases in the last 33 years, these bed reductions are proportionally higher – "a drop of 42% in the number of acute beds per 1000 population".

- Changes in medical care have shortened length of stay in hospital from 9.4 days for acute cases in 1979 to about 3 days in 2011.

- Policies such as Care in the Community have changed bed stock taken up by mental health, learning disability and geriatric services from around 54% to 30% in just over 30 years.

- Daily bed occupancy across all hospitals reached over 90% on several days in England in 2011-2012. "Such high occupancy rates reduce the time available for cleaning between patients and increase the chances of infection." Yuk.

The new UK antimicrobial resistance strategy and action plan

BMJ 2013;346:f1601

Anthony S Kessel, honorary professor, Mike Sharland, professor in paediatric infectious diseases

- Here's another bloody thing to worry about.
- The rise in antimicrobial resistance (AMR) poses a threat to healthcare in the UK.
- The rise of AMR as a serious health threat is due to the international spread of multidrug resistant (MDR) gram negative bacteria, the global overuse of antibiotics in humans and animals, and the almost complete lack of new antibiotic development.
- There has been an 85% reduction in rates of MRSA bloodstream infections in England between 2003 and 2011. MRSA is responsible for only 2% of bloodstream infections in England now.
- Gram negative organisms, particularly E. Coli, comprise nearly half of the 100,000 annual blood stream infections in England.
- Klebsiella is an important pathogen in the spread of resistance. Many EU countries are now reporting Klebsiella MDR rates of 25-40%.
- High global rates of MDR gram negative infections have led to "a rapid rise in the use of carbapenem antibiotics... as an empirical treatment of suspected sepsis." This has led in turn to a "rapid increase in carbapenemare producing organisms".
- Only one or two new antibiotics that target gram negative organisms are likely to be marketed in the next decade, raising concern that there will be "virtually untreatable infections" which will threaten NHS care.
- The new UK strategy recognises that AMR, infection prevention and control and antibiotic stewardship "are closely interconnected and all need to be strengthened".

- Antimicrobial prescribing needs to be evidence based and efficiently targeted. For primary care there is a TARGET antimicrobial toolkit.
- Drug-bug combination resistance rates may result in national outcome measures.
- It's the end of the bloody world, that's what it is.

Wind turbine noise

BMJ 2012;344:e1527

Christopher D Hanning, honorary consultant in sleep medicine, Alun Evans, professor emeritus

- The evidence of a link between adequate sleep and human health is "overwhelming".
- Shortly after wind turbines started to be erected, complaints emerged of a degree and consistency to "constitute epidemiological evidence of a strong link between turbine noise, ill health and disruption of sleep".
- Wind turbines produce dynamo gear train noise and aerodynamic blade noise through an air stream which constantly changes in velocity and direction, especially at night.
- The noise is described as "impulsive noise" (swishing or thumping) and is "much more annoying than other sources of environmental noise".
- Noise levels and setback distances have not been revised in the UK since 1997.
- Turbines must be set back from human habitation by a minimum distance of 350-500m.
- Wind turbine noise has a large low frequency and infrasound component that is "attenuated less with distance than high frequency noise". Current measurement techniques obscure the contribution of low frequenty and infrasound.
- Low frequency and infrasound noise in lab studies was found to be more annoyng than higher frequency noise and caused nausea, headaches, poor sleep and cognitive and psychological impairment.
- In a survery of people living in the vicinity of 2 US wind farms, those living within 375-1400m reported worse sleep and more daytime sleepiness, and had lower mental health component scores on the short form 36 health survey, than those who lived 3-6.6 km from a turbine.

- A New Zealand study showed lower "health related quality of life" in people who lived less than 2 km from turbines.

Low carbohydrate-high protein diets

BMJ 2012;344:e3801

Anna Floegel, nutritional epidemiologist, Tobias Pischon, professor

- Low carbohydrate-high protein diets (such as the Atkins diet) have been suggested to have health benefits over low fat diets, mainly on the basis of short term studies.
- The suggested benefits include reduction in triglyceride and HbA1c concentrations and reduction in systolic blood pressure.
- However the long term health effects are unclear, particularly as adherence to these diets is associated with higher cardiovascular mortality in prospective cohort studies.
- A 15 year study of over 43,000 Swedish women using food frequency questionnaires showed that women on low carbohydrate-high protein diets had a higher incidence of cardiovascular disease in a dose-response manner. Those women in the highest category of low carbohydrate-high protein diet had a 62% higher incidence of cardiovascular disease.
- A low carbohydrate-high protein diet is likely to be low in fibre, vitamins and minerals and high in iron, cholesterol and saturated fat.
- The increased cardiovascular risk with these diets "was seen consistently across different protein sources and when adjusted for fat quality".
- A recent randomised controlled trial showed that a low glycaemic index and low protein diet improved cardiovascular disease risk markers in addition to reducing body weight.

Chronic kidney disease controversy: how expanding definitions are unnecessarily labelling many people as diseased

BMJ 2013;347:f4298

Ray Moynihan, senior research fellow, Richard Glassock, emeritus professor, department of medicine, Jenny Doust, professor

- In 2002 the United States Kidney Foundation launched a framework classifying chronic kidney disease – to considerable controversy.
- The framework uses the term "chronic kidney disease" to include contitions that affect the kidney and may result in progressive decrease in kidney function, or complications from decreased function.
- CKD was defined as kidney damage or decreased function for 3 months or more.
- An eGFR less than 60 was arbitrarily adopted, and similarly albumin/creatinine ratio greater than or equal to 3.
- Initially there were 5 stages of CKD, but in 2012 stage 3 was split into 3A, 3B and three extended categories for persistent albuminuria.
- A key rationale for the definition was that decreased eGFR and albuminuria are associated with increased risk of death and end stage renal failure. A meta-analysis showed that reduced eGFR and albuminuria were "consistently associated with cardiovascular mortality".
- The assumption was made (in 2002) that early identification could prevent progression (especially in "severe forms of specific kidney diseases").
- In 2012, the National Kidney Foundation said that the claims about early detection "remained to be proven".
- The US Preventive Services Task Force found "insufficient evidence" for general population screening and advised there were no studies on the benefits of early CKD treatment in patients without diabetes or hypertension.

- The US Preventive Services Task Force are a pretty tough-sounding organisation: think combat fatigues, cigars, crew-cuts and machine guns. Those are good guys to have on your side if this debate turns nasty.
- "Nine of the 16 working group members who produced the 2012 guidelines declared financial ties with drug or device companies." Hm. Strong smell of rat.
- The threshold eGFR chosen to define disease was set at 60ml/min/1.73m2, "about half that of a normal level of a young adult". This definition has resulted in almost 14% of US adults being labelled as having CKD, and as many as 1 in 6 in Australia.
- At least 1 in 3 of people who meet the definition of CKD are classified as stage 3A; most of them are older than 65 and many will have an eGFR in the 5th to 95th percentile for their age. About 3 in 4 of these have no albuminuria.
- Kaiser Permanente have adapted the framework to take age into account, reducing the prevalence of CKD to 3% compared with 14% arising from the current framework definition.
- Kaiser Permanente – that's another tough-sounding outfit. Spiked metal helmets, monocles, big moustaches. Famous for their relentless discipline and trench-digging expertise.
- Half of people aged 70 or above are labelled as having, or at risk of, CKD. Dutch researchers consider that eGFR of 60 is "normal" for men over 60 and women over 50 years old, and "cannot be used to define a diseased population".
- Although 1 in 8 adults are being labelled as CKD on 1 in 3000-5000 are newly treated for end stage renal disease each year.
- A Norwegian study surveyed 65,000 people and found that less than 1% of people with an eGFR of 45-59 went on to develop end stage renal disease after 8 years of followup.
- Many patients labelled as having CKD "may already warrant treatment... regardless of CKD status" (ie they're already on what would be the right treatment for CKD, but they're on it for other conditions).
- Labelling so many low risk patients as CKD is like "a fishing trawler: it captures many more innocent subjects than it should".
- A qualitative research study of GPs and practice nurses found that nearly all had "reservations as to whether CKD was really a

disease", and some expressed reservations about medicalisation causing harm.

- There remain questions about the association between CKD and cardiovascular disease and whether this "adds meaningfully to the traditional assessment of risk".
- Problems with the precision and reliability of eGFR measurements remain and periods longer than 3 months may be needed to make a "diagnosis". A Norwegian study suggested that an abnormal eGFR should persist for 12 months to qualify – this could reduce the prevalence of stage 3 disease by 37%.
- There is uncertainty about what level of albuminuria constitutes "meaningful" kidney damage. An albumin/creatinine ration of 3-30mg/mmol "is not pathognomonic of persisting kidney damage". It can be transitory and affected by extraneous factors (such as a poached egg strolling by outside the lab window, whistling nonchalantly, just as the analysis is made).
- The US Preventive Services Task Force advise that the benefits of treating people at risk of cardiovascular disease through any screening programme for CKD must be weighed against the risk of drug side effects.
- Studies of hypertension suggest that disease labelling could cause psychological distress and absenteeism from work whilst decreasing quality of life.
- The introduction of the framework has increased referrals for CKD, with nephrology referrals up 60% in one UK trust.
- The authors advise that a review should be conducted "with minimal conflicts of interest".
- It is in everyone's interest to maximise prevention of CKD and minimise the risks and costs of overdiagnosis.
- The real question is this: in a straight fight between the US Preventive Services Task Force and Kaiser Permanente, who would you put your money on?

QUIZ NO 6

Epistaxis: If someone complains of nose bleeds, the first line of advice is "Stop picking your nose".	☐ True ☐ False
Nasopharyngeal angiofibromas usually occur in posh elderly women, who get them because they wear pince-nez glasses to watch Downton Abbey.	☐ True ☐ False
If nose bleeds are associated with face or ear pain, something serious may be going on.	☐ True ☐ False
If the patient history suggests anterior epistaxis and no vessel is seen, petroleum jelly can be tried.	☐ True ☐ False
Dietary fats and breast cancer risk: It is clear that there are no environmental, modifiable breast cancer risk factors.	☐ True ☐ False
The role of dietary fat in breast cancer is clear.	☐ True ☐ False
Zhong et al looked at the association between eating fish and n-3 PUFAs.	☐ True ☐ False
Zheng found that n-3 PUFAs resulted in a causal reduction in breast cancer risk of 14% in high PUFA eaters.	☐ True ☐ False
However, nobody wants to be labelled a "high PUFA eater".	☐ True ☐ False
Zheng found no association with fish-eating, but fatty fish is the main source of marine n-3 PUFAs.	☐ True ☐ False
Action plans for patients with COPD: Patients with repeated exacerbations of COPD often get frequent emergency admissions to the hospital.	☐ True ☐ False
If people have proper COPD action plans they won't get scared out of hours and rush off to hospital any more.	☐ True ☐ False

On the other hand, perhaps they will.	☐ True ☐ False
"Infection stones" are the commonest type of stone.	☐ True ☐ False
Management of renal colic: Women are more prone to renal stones than men.	☐ True ☐ False
"Renal colic" may be caused by lymphadenopathy, blood clots, sickle cell disease, diabetes or long term analgesics.	☐ True ☐ False
Patients with high vascular risk and left sided renal colic can safely be looked after at home (check with MPS/MDU).	☐ True ☐ False
An obstructed infected kidney should be treated at home with large doses of cefalosporins, quinolones and a mustard poultice.	☐ True ☐ False
When it comes to acute flank pain, a clinician's "gut feeling" is as good as an NCCT and considerably cheaper.	☐ True ☐ False
Pethidine is more effective and less likely to cause side effects than NSAIDs in cases of renal colic.	☐ True ☐ False
Alpha 1 blockers and calcium antagonists have no effect on the passage of stones.	☐ True ☐ False
This is one area where there may actually be some mileage in your Mum's conviction that almost anything will get better if you put a hot water bottle on it.	☐ True ☐ False
Updated guidance on familial breast cancer: About 20% of all breast cancers are attributed to mutations of BRCA1, BRCA2 and TP53 genes.	☐ True ☐ False
It is easy to remember the guidelines for referral of patients with no personal history of breast cancer but with relatives who've had breast cancer.	☐ True ☐ False
A first degree female relative is the first female in	☐ True

your family to go to university and get a degree.	☐ False
Secondary care will use a "probability calculation method" to evaluate referred patients.	☐ True ☐ False
Mammography is more cost effective than MRI for high risk women with known BRCA or TP53 mutations.	☐ True ☐ False
Tamoxifen may be offered to high-risk women pre- or post-menopause.	☐ True ☐ False
Secondary care may discuss mastectomy or BSO with women who have known gene mutations (BRCA or TP53).	☐ True ☐ False
The new referral guidelines will reduce referral rates to specialist genetic clinics.	☐ True ☐ False
Should GPs be fined for rises in avoidable emergency admissions to hospital? - Yes: Collaborative working across the continuum of health and social care may reduce admissions.	☐ True ☐ False
GPs' behaviours and attitudes can be influenced by "enablers, levers and incentives" (carrots and sticks).	☐ True ☐ False
Should GPs be fined for rises in avoicable emergency admissions to hospital? - No: Emergency hospital admissions have risen by 37% in the last 10 years.	☐ True ☐ False
Incentives paid to GPs in the time of PBCs (2005-11) did not reduce emergency admissions.	☐ True ☐ False
Risk stratification and case management have been proven to reduce emergency admissions.	☐ True ☐ False
In small GP practices admission rates are unlikely to be affected by chance or volatility.	☐ True ☐ False
Two recent RCTs of interventions designed to keep people out of hospital showed increased deaths in intervention groups.	☐ True ☐ False
Dr Hairy's handling of issues like this one is	☐ True

scrupulously even-handed.	☐ False
The hospital bed - on its way out?: The number of acute care beds in the NHS has fallen by 35% in the last 33 years.	☐ True ☐ False
The reduction in acute beds since 1979 is equivalent to a 42% drop in beds/1000 population.	☐ True ☐ False
There is now so little time to clean up between patients that it might be advisable not to taste that "complimentary chocolate" on your pillow.	☐ True ☐ False
The new UK antimicrobial resistance strategy and action plan: Antimicrobial resistance (AMR) is due to the spread of multidrug resistant (MDR) gram negative bacteria, overuse of antibiotics and lack of new antibiotic development.	☐ True ☐ False
The new UK AMR strategy only relates to secondary care.	☐ True ☐ False
Antimicrobial prescribing needs to be evidence based and targeted efficiently.	☐ True ☐ False
There is a primary care antimicrobial toolkit called Bug-off.	☐ True ☐ False
Wind turbine noise: The noise from wind turbines is "impulsive noise", which means it can be jolly annoying	☐ True ☐ False
All that swishing and thumping - it's like living next to the guillotine during the French Revolution, but without any of the fun.	☐ True ☐ False
Wind turbine noise, especially infrasound noise, is adequately measured by current techniques.	☐ True ☐ False
The current set back distance of human habitation from wind turbines (350-500 metres) seems to be fine.	☐ True ☐ False
People's health and quality of life may be affected by wind turbines sites up to 2km from their homes.	☐ True ☐ False

Swishing and thumping sounds may also be heard at some nuclear reactors - for example the swishing of a tidal wave overwhelming the station, followed by a loud thump as the core explodes.	☐ True ☐ False
Low carbohydrate-high protein diets: The suggested benefits of low carbohydrate-high proteins diets have been confirmed by long-term studies.	☐ True ☐ False
Long term adherence to low carbohydrate-high protein diets is associated with a higher incidence of cardiovascular disease.	☐ True ☐ False
Life isn't worth living without the occasional baked potato.	☐ True ☐ False
Chronic kidney disease: CKD is defined as kidney damage or decreased function lasting for 3 months or more.	☐ True ☐ False
It is absolutely clear that early detection of CKD prevents progression.	☐ True ☐ False
There are no studies on the benefits of early treatment of CKD in patients without diabetes or hypertension.	☐ True ☐ False
Of the 16 working group members who produced the 2012 guidelines, how many declared financial ties with drug or device companies?	☐ None ☐ Nine
The threshold eGFR chosen to define the disease results in about 14% of US adults and 1/6 of Australians being labelled as having CKD	☐ True ☐ False
Dutch researchers say that an eGFR of 60 is normal in men of 60 and women over 50.	☐ True ☐ False
Mind you, those Dutch researchers are so laid-back they'll say anything's normal. If you ask me they're all stoned.	☐ True ☐ False
Only a tiny proportion of people with CKD are stage 3A, and most of those are young and have proteinuria.	☐ True ☐ False

CKD prevalence could be reduced from 14% to 3% if the Kaiser Permanente framework was adopted.	☐ True ☐ False
In Norway they say "No way" to the 3 month period for diagnosis, only labelling patients as CKD after a 12 month period of abnormal eGFR. This makes stage 3 prevalence in Norway much more afjordable!	☐ Top joke ☐ Oh dear ☐ Points for hilarity: Norse out of Sven

Making a diagnosis in patients who present with vertigo

BMJ 2012;345:e5809

D Kaski, neurology registrar, A M Bronstein, professor of neuro-otology

- Vertigo "is the illusion of movement" whereas labyrinthitis is an inflammation in the labyrinth and is rare.
- Vestibular neuritis (inflammation of the vestibular nerve) is more common than labyrinthitis, with an incidence of 3.5/100,000 population.
- In a study of patients referred to a neuro-otology service with vestibular neuritis mislabelled as labyrinthitis, "the diagnosis was correct in only 15%; most of the others had benign paroxysmal positional vertigo (BPPV) or vestibular migraine".
- Vestibular neuritis presents as a single acute attack of continuous rotational vertigo, nausea (and vomiting) and imbalance. The symptoms last for several days even when the head is still.
- Vertigo in BPPV is induced by head movements and is diagnosed by the Hallpike manoeuvre.
- The Hallpike manoeuvre consists of taking a patient into your front hall and suddenly shouting "Look at that!", pointing to a stuffed pike you caught on a fishing expedition in 1975. If the patient gets all dizzy and falls down, it's vertigo.
- Recurrent episodes of BPPV are interrupted by periods of relief.
- Patients with vestibular neuritis can remain upright using "furniture walking" but "patients with cerebellar stroke are unable to stand".
- Patients with vestibular neuritis have nystagmus when looking forward. The nystagmus is unidirectional.
- "In the absence of other central nervous system symptoms or signs", unidirectional nystagmus points to a peripheral rather than central problem.

- Patients with vestibular neuritis have unilateral loss of the vestibulo-ocular reflex which normally moves the eyes in the opposite direction to head movement.
- Red flags associated with "acute dizziness" include unilateral hearing loss, abnormal neurological signs or symptoms, and new headache.

Irritable bowel syndrome

BMJ 2012;345:e5836

Alexander C Ford, senior lecturer and honorary consultant gastroenterologist, Nicholas J Talley, professor of medicine

- IBS is a chronic functional disorder of the lower gastro-intestinal tract, characterised by pain "in association with" an alteration in stool form or frequency.
- Abdominal bloating may occur, and pain may be relieved by defaecation.
- The Rome criteria classifies IBS into diarrhoea predominant (IBS-D), constipation predominant (IBS-C), or mixed (IBS-M).
- IBS-D is the commonest subtype and IBS-M is the least common.
- IBS is more common in people under the age of 50.
- It is more common in women.
- IBS is associated with psychiatric illnesses and "maladaptive coping strategies".
- Small bowel and colonic transit times are abnormal in IBS.
- Gas production, visceral hypersensitivity and pain processing may account for the pain of IBS.
- "Perturbations" of gastrointestinal flora and chronic low grade inflammation have been implicated as possible causes.
- "In a patient without lower gastrointestinal alarm symptoms who has longstanding typical symptoms of IBS" the diagnosis should be made on clinical grounds.
- Rome III criteria are the "current accepted diagnostic standard" used by gastroenterologists for diagnosing IBS. However, the criteria "have not been validated extensively".
- Routine blood tests such as FBC and CRP are often done, and coeliac serology is worthwhile as coeliac disease in patients with suspected IBS is around 5%.
- In a meta-analysis of 6 trials with an NNT of 6, ispaghula improved symptoms when compared with placebo.

- A study of food elimination based on the results of IgG testing showed lower symptom scores in those on elimination diets compared with "sham" diets.
- FODMAPs diet stands for "fermentable oligosaccharides, disaccharides, monosaccharides and polyols". It may have effects on fermentation and osmosis. Its main reason for its existence, however, is that it sounds a bit like "foodmap" and we like to keep the acronym-count high.
- High FODMAP foods include apples, cherries, peaches/nectarines, artificial sweeteners, most lactose containing foods, legumes, broccoli, brussel sprouts, cabbage and peas. Bang goes Christmas lunch.
- A crossover trial showed that symptoms were worse on a high FODMAP diet.
- Another trial suggested that a gluten free diet might benefit people with IBS who test negative for coeliac disease.
- Exercise may be effective for IBS (and CFS and fibromyalgia).
- Most of the above trials were in secondary or tertiary care.
- A tiny trial (RCT) in tertiary care "suggests" that melatonin may help women with IBS.
- Placebo response rate in IBS (meta-analysis) is about 40%.
- A really teeny-weeny trial (RTWT) which also turned out to be rather ill-advised (RIA) showed that trying the Hallpike manoeuvre on a patient with vertigo + IBS may lead to a carpet cleaning bill of up to £250 (CCB=250).
- Antispasmodics inhibit smooth muscle contraction via competition with ACh at parasymptomatic nerve endings.
- Peppermint oil causes smooth muscle relaxation via calcium channel blockade.
- Efficacy of antispasmodics has been confirmed in 2 recent meta-analyses. NNT = 5.
- Antispasmodic studies included otilonium, cimetropium, hyoscine, pinaverium and dicycloverine. All were more effective than placebo, but the NNH (number needed to harm) was 17.5.
- NNT with peppermint oil was 2.5. Adverse reactions were "rare".

- Two meta-analyses suggest that antidepressants and SSRIs (selective serotonin reuptake inhibitors) were more effective than placebo for IBS (no primary care trials).
- NNT for 9 RCTs using tricyclics gave a "pooled" NNT of 4, with similar NNT for SSRIs (2.5).
- Alosetron is a 5-HT receptor antagonist licensed in the US for women with severe IBS-D. Meta-analysis shows an NNT of 8, but it is associated with ischaemic colitis and severe constipaton.
- Procalopride is a 5-HT agonist and is effective for chronic idiopathic constipation with an NNT of 6. Theoretically it may benefit IBS-C.
- A non-absorbable antibiotic rifamixin has been tested in 2 large placebo-controlled trials (2 weeks' treatment and followup at 12 weeks). The result showed significant reduction in symptoms.
- Some probiotics may have "anti-inflammatory properties or ameliorate visceral hypersensitivity". Pooled data from trials showed an NNT of 4. A "trend" was seen towards bifidobacteria improving symptoms. Adverse effects were rare.
- Psychological disorders are two to threefold higher in patients with IBS. Which one is cause and which is effect? As my Mum used to say, "If you spend too long on that toilet, you'll end up round the bend."
- A Cochrane review suggested that CBT, hypnotherapy and psychotherapy were marginally better than no treatment. A second meta-analysis showed a pooled NNT of 4 for these therapies.
- A hypnotherapy trial in tertiary care showed "significant benefit" (over other interventions). This was one of the largest trials of IBS to date.
- A mindfulness training trial showed "significant" reduction of IBS symptoms.
- Acupuncture in China was more effective than drug therapy (meta-analysis).
- Iberogast (also known as STW5), a combination of plant extracts, shows "superiority" over placebo.
- Lubiprostone and linaclotride are effective in treating chronic idiopathic constipation and may be therapeutic in IBS-C.

Necrotising fasciitis

BMJ 2012;345:e4274

Helen Yasmin Sultan, consultant, Adrian A Boyle, consultant, Nicholas Sheppard, specialist registrar

- Necrotising fasciitis is also known as gas gangrene, Fournier's gangrene and "necrotising soft tissue infection".
- 500 cases occur in the UK per year.
- Even with surgery, mortality is 20-40%.
- Delay in diagnosis increases mortality.
- However, it is very difficult to diagnose.
- Initial symptoms are non-specific until rapid deterioration, septicaemia and shock.
- Fever and pain develop first, the pain being disproportionate to the clinical findings. Cellulitic changes may then occur and may mimic other inflammatory conditions in appearance or symptoms (haematoma, bursitis, septic arthritis, DVT...)
- The haemorrhagic bullae, crepitus and skin necrosis may occur from day 5 onwards.
- Some patients present with severe pain, without fever, and appear well. They may have exquisite pain and tenderness and overlying sensory loss.
- Pain may prevent weight bearing or limb movement.
- IV drug users often present with no systemic symptoms.
- It is more common in people with diabetes, chronic hepatitis or malignancy, and in people who inject drugs.
- Varicella infection is a recognised risk factor in children.
- 25% of patients have no history of co-morbidity or trauma.
- The classic cyanotic and bullous skin changes may only appear late in the process; "however the site of the infection may appear unusual". (That's a big help, isn't it?)
- The patient may be well initially but will deteriorate "despite treatment with antibiotics".

- Patients with severe pain without trauma may have necrotising fasciitis. A search for "covert sepsis" and checking ESR and CRP "is advisable".
- Hyponatraemia, sepsis and signs of soft tissue infection are "highly suspicious" of necrotising fasciitis.
- CT and MRI are "sensitive"; ultrasound is tricky; x-ray may show subcutaneous gas.
- The mainstay of investigation and treatment is surgical exploration. ("If in doubt, poke about.")
- Necrotising soft tissue infections are caused by "mixed pathogens, including gas forming bacteria such as clostridium species".
- The average hospital stay is 33 days, and the patient will then need referral to a plastic surgery centre.

Post-traumatic stress disorder

BMJ 2012;344:e3790

Ruth V Reed, specialty registrar in child and adolescent psychiatry, Mina Fazel, NIHR postdoctoral research fellow, Lorna Goldring, general practitioner

- "Post-traumatic stress disorder (PTSD) is a severe, prolonged, and impairing psychological reaction to a distressing event".
- The event must be "exceptionally threatening or catastrophic" – sexual violence is a particularly potent cause.
- Serious illnesses or medical interventions can cause PTSD.
- Intrusive imagery of past events, bodily re-experiencing, nightmares, flashbacks, irritability and insomnia occur.
- Sufferers may have difficulty remembering aspects of the event and avoid reminders of it.
- Adolescents with PTSD may have aggressive or withdrawn behaviour and find it difficult to relate to their peers.
- The prevalence may be underestimated by GPs and it may be underdiagnosed.
- 80% of PTSD cases "are co-morbid" with depression, panic, substance misuse and personality disorders. These other disorders may be considered to be the primary diagnosis.
- Ask about "re-experiencing" symptoms, as these may not be volunteered by patients due to "shame or distress".
- Culture and language are barriers to diagnosis.
- Intrusive imagery and auditory re-experiencing can be misinterpreted as psychotic symptoms.
- Most treatments for anxiety and depression have limited efficacy in PTSD, and patients may be assumed to have treatment-resistant anxiety or mood disorder.
- Some patients settle without treatment in the first year, but others develop chronic symptoms and are at risk of suicide.
- Unemployment and marital separation can arise.

- Ask about "potentially traumatic events" in patients with "treatment resistant mood or anxiety disorders, or unexplained physical symptoms".
- Trauma-focussed CBT or "eye movement desensitisation and reprocessing" are considered to be helpful (NICE and other "major guidelines"). Otherwise, paroxetine or mirtazepine "may be offered" in primary care.

Diagnosis and management of peripheral arterial disease

BMJ 2012;345:e5208

G Peach, specialist registrar in vascular surgery, M Griffin, academic foundation year 2 trainee, K G Jones, consultant vascular surgeon, M M Thompson, professor of vascular surgery, R J Hinchliffe, Higher Education Funding Council for England clinical senior lecturer and consultant vascular surgeon

- Peripheral arterial disease (PAD) has a prevalence of 15-20% in people over 70.
- Critical limb ischaemia costs the NHS £200 million a year (500-1000 new cases).
- PAD has modifiable and non-modifiable risk factors, and patients with PAD have a six-fold higher risk of death from cardio-vascular disease.
- Smoking is the most important modifiable risk factor.
- Failure rate of surgical bypass grafts is 3 times higher in those who continue to smoke.
- In diabetics, "a meta-analysis found that a 1% increase in HbA1c is associated with a 26% increase in the risk of developing PAD".
- "Patients with diabetes and foot 'ulcers' should be seen by a multidisciplinary foot team" (ideally within 24 hours) (which would be impossible to organise in real life).
- A BP greater than 160/95 "increased the risk of developing intermittent claudication 2.5 fold in men and fourfold in women".
- Intermittent claudication in the calf is related to superficial femoral artery PAD.
- Iliac PAD causes buttock pain; common femoral PAD caused thigh pain; tibio-peroneal PAD causes foot pain.
- Rapid exacerbation of symptoms may indicate new arterial occlusion, usually caused by plaque rupture, eg. in a popliteal aneurysm or heart emboli.

- Critical limb ischaemia causes ulceration, gangrene or rest pain for "more than 2 weeks".
- True rest pain "usually" affects the toes or foot.
- Only 20-25% of patients with claudication are likely to deteriorate. Only 1-3% of patients with intermittent claudication will need amputation in a 5 year period.
- 12% of patients with critical limb ischaemia need amputation within 3 months and up to 25% will die within a year. The five year survival rate is about 50%.
- Pallor on limb elevation and hyperaemia when the leg is dependent (Buerger's sign) is a sign of PAD (loss of capillary autoregulation).
- "If the presence of pulses is in doubt, the ankle bracial pressure index (ABPI) can be measured" with a Doppler machine.
- ABPI = highest ankle pressure/highest arm pressure (right or left).
  - >1.2 may mean "heavy vessel calcification"
  - 0.9 – 1.2 = normal
  - 0.5 – 0.9 = PAD
  - <0.5 = critical limb ischaemia
- Refer
  - Patients with claudication (but not if unaffected quality of life)
  - Patients with tissue loss or sudden deterioration of symptoms
  - New onset diabetic foot ulcers
- Secondary care investigation may include duplex US scan, digital subtraction angiography, or (better) CT angiography/MRI angiography.
- Consider thrombophilia screen and serum homocysteine in patients under 50 with PAD.
- (Tell them to give up smoking.)
- ACE inhibitors reduce the risk of cardiovascular events in patinets with PAD, and these "are generally recommended" as first line for hypertension. However... 24% of patients with PAD have "co-existent renal artery stenosis" (NNH = 4).

- Clopidogrel is now recommended by NICE as first line antiplatelet treatment for patients with PAD (better, safer and cheap).
- Exercise "can improve walking distance" up to 200%.
- "Structured exercise programmes may improve walking distance as effectively as angioplasty".
- An increasing proportion of PAD lesions can be treated endovascularly.
- Angioplasty or bypass for severe ischaemia have similar results, "although angioplasy had significantly higher rates of reintervention" (28% vs 17%).
- "When revascularisation is unlikely to be successful or if patients have comorbidities that might prevent them making use of a salvageable limb, primary amputation may offer the best quality of life".

Myasthenia gravis

BMJ 2012;345:e8497

J Spillane, clinical research associate, E Higham, general practitioner partner, D M Kullmann, professor of neurology

- Myasthenia gravis is an autoimmune disorder of neuromuscular transmission characterised by fatigable muscle weakness.
- It is mediated by antibodies against the postsynaptic acetylcholine receptor or muscle specific tyrosine kinase.
- About 10% of patients with myasthenia gravis have a thymoma.
- Patients typically present with ocular symptoms, typically diplopia or ptosis.
- Weakness then becomes generalised in 80% of patients.
- The most serious complication is acute respiratory failure (myasthenic crisis).
- Myasthenia gravis may be under-recognised, particularly in the elderly, because symptoms such as dysphagia, fatigue and slurred speech can have a broad differential diagnosis and ptosis may be considered to be age-related (or aponeurotic).
- Some patients have negative serology and normal electrophysiology, so "a thorough history is central to the diagnosis".
- Muscle weakness is worse at the end of the day and may worsen with heat, infection, pregnancy, stress, surgery, menstruation or post-partum.
- Diplopia and ptosis are the most common presenting symptoms but 80% of people go on to develop generalised weakness.
- Weakness is more typically in the upper limbs than lower limbs.
- Bulbar weakness can cause difficulty chewing, slurred speech or dysphagia, choking or nasal regurgitation of liquids.
- Respiratory weakness may cause exertional dyspnoea or orthopnoea.
- Look for weakness of forced eyelid or mouth closure.

- Look for development of ptosis during sustained upgaze. (Try the Hallpike manoeuvre again, this time with the pike fastened to your hall ceiling.)
- Test shoulder abduction before and after repeated arm movements.
- The ice test distinguishes myasthenia gravis from other forms of ptosis. Crushed ice in a latex glove is applied to the eye for 3 minutes, and in myasthenia this leads to improvement of ptosis. In other conditions it just leads to a very chilly eye.
- You can use the glove full of crushed ice to treat the patient's haemorrhoids afterwards.
- About 85% of patients with generalised myasthenia gravis will have antibodies to acetylcholine receptors and 40-70% of the rest are positive for mucle specific tyrosine kinase. Some patients, however, will be persistently seronegative.
- Specialised electrophysiological tests may help confirm the diagnosis but the sensitivity is only 70% in repetitive nerve stimulation and is of low specificity in single fibre electromyography..
- The edrophonium (Tensilon) test is rarely done as it can cause life threatening bradycardia.
- CT of the thorax is "required in all patients" with diagnosed myasthenia gravis to exclude a thymoma.
- Pyridostigmine is first line treatment. Oral steroids are generally first line immunosuppressive treatment, but may need to be initiated in hospital as initial high doses may exacerbate weakness.
- Other immunosuppressive drugs may be used second line. Rituximab may be used in severe resistant myasthenia gravis and immunoglobulin infusion and plasma exchange may be used for myasthenic crisis.
- Thymectomy "should always be performed if a thymoma is suspected".
- Thymectomy "is not generally carried out" in patients with muscle speclific tyrosine kinase antibodies, in last onset myasthenia gravis or purely ocular disease.

- Aminoglycoside and quinolone antibiotics, quinine and IV magnesium are contraindicated in myasthenia gravis, as they can impair neuromuscular transmission.

Rotavirus vaccine: a welcome addition to the immunisation schedule in the UK

BMJ 2013;346:f2347

Miren Iturriza-Gómara, Wellcome Trust tenure track fellow, Nigel Cunliffe, professor of medical microbiology

- Live, attenuated, two dose, oral monovalent vaccine (Rotarix) will be given with other routine vaccinations to children by the age of 4 months.
- Rotavirus is the most common cause of severe gastroenteritis in infants and children, and worldwide it causes about 453,000 deaths a year in children under 5 years (>90% of these deaths occurring in developing countries).
- In the UK, rotavirus results in about 750,000 episodes of diarrhoea and 80,000 GP consultations a year. It accounts for 45% of hospital admission for gastroenteritis in children under 5 years.
- Clinical trials in middle and high income countries showed >85% vaccine efficacy against severe rotavirus gastroenteritis. Post vaccination surveillance has shown reductions in mortality in middle income Latin American countries.
- Rotavirus vaccination reduces rotavirus infection in unvaccinated people (herd protection).
- Although there is no clear relationship between rotavirus vaccination and intussusception, "a low rate of intussusception" was reported after Rotarix vaccination in Mexico and Australia.

What's this? An approving Dr Hairy writeup for a new preventive treatment programme? No scepticism? No curmudgeonly remarks about the pharmaceutical companies and their profits? What's going on? Oh, but wait a minute...

Cytisine, the world's oldest smoking cessation aid

BMJ 2013;347:f5198

Judith J Prochaska, associate professor of medicine, Smita Das, resident physician, Neal L Benowitz, professor of medicine and bioengineering and therapeutic sciences

- Cytisine is an alkaloid derived from the plant cytisus laburnum and has been used as a smoking cessation aid in eastern and central Europe for nearly 50 years. But you can't use it in the West thanks to a combination of bureaucracy and drug company self-interest.

- It was discovered in 1818, isolated in 1865, and its actions were documented as "qualitatively indistinguishable from that of nicotine" in 1912.

- It was brought to the market in 1964 under the brand name Tabex and is now produced by a Bulgarian drug company.

- Clinical trials on cytisine's benefits in smoking cessation were published in 1960s and 1970s in eastern European journals. Quit rates ranged from 41% - 65% at the end of treatment, but the trials' methods "did not meet Western regulatory standards".

- A 2013 meta-analysis in Thorax estimated the efficacy of cytisine to be "significant" and comparable to published effects for nicotine replacement, bupronion [and others] with a relative rask of abstinence of 1.57. A Cochrane review showed even higher relative risk of abstinence.

- Regulatory approval in the West would require "serious investment" (upward of £0.64 bn). The drug industry is unlikely to finance research because of the high cost-profit ratio. Not to mention the fact that they're already making money hand over fist from far-more-expensive products of their own.

- Cytisine is half to a twentieth of the cost of other cessation drugs.

- Tabex can be bought on the internet – but beware of poor quality and counterfeit formulations! However, you'll know if you've got the right stuff, because you get a free Bulgarian bride with every box.

Management of autism in children and young people: summary of NICE and SCIE guidance

BMJ 2013;347:f4865

Tim Kendall, director, consultant psychiatrist and medical director, professor, Odette Megnin-Viggars, systematic reviewer, Nick Gould, consultant, emeritus professor, professor, Clare Taylor, senior editor, Lucy R Burt, research assistant, Gillian Baird, consultant paediatrician, professor of paediatric neurodisability, on behalf of the Guideline Development Group

- Autism occurs in 1% of children and young people.
- The core features are persistent impairments in reciprocal social interaction and communication, and restricted, repetitive patterns of behaviour, interests or activities (with or without intellectual disability).
- 70% of children with autism have mental and behavioural disorders. Sensory sensitivities, constipation and sleeping and eating problems also occur.
- Children with autism should have access to specialist community based multidisciplinary teams, including health, mental health, learning disability, education and social services.
- Common co-existing problems include anxiety and depression, epilepsy, and ADHD.
- Interventions for the core features of autism include play-based strategies to encourage joint attention and reciprocal communication. The intervention should be delivered by trained professionals.
- Antipsychotics, antidepressants, anticonvulsants and exclusion diets should not be used for core features of autism.
- Secretin, chelation and hyperbarix oxygen should not be used in any circumstances for autism, as there is lack of evidence of effectiveness and clear evidence of harm.

- If there is no mental health issue, behavioural problem, physical disorder or environmental problem triggering "behaviour that challenges", offer a "psychosocial intervention". Antipsychotic medication may be considered if psychosocial intervention cannot be delivered.
- Offer parents, siblings and/or carers an assessment of their own needs.
- For young people aged 16 or over whose needs are complex or severe, use the care programme approach to coordinate their needs.
- Inform young people and the carers about adults services and the right to a social care assessment at age 18.

Non-steroidal anti-inflammatory drugs (NSAIDs)

BMJ 2013;346:f3195

Richard O Day, professor of clinical pharmacology, Garry G Graham, professorial visiting fellow

- NSAIDs act by inhibiting COX-1 and COX-2 enzymes.
- "There are two broad groups of NSAIDs—the older, traditional, non-selective NSAIDs that inhibit both COX-1 and COX-2 and the newer, selective COX-2 inhibitors that predominantly inhibit COX-2."
- There is little difference in efficacy of NSAIDs but patients vary in their responses to different NSAIDs.
- Both groups of NSAIDs are effective for acute gout. Colchicine and corticosteroids are alternatives.
- Both groups of NSAIDs are used for osteoarthritis and low back pain, but "their effectiveness is generally small".
- A systematic review showed that walking and quadriceps strengthening decrease pain and disability from osteoarthritis of the knee.
- Topical NSAIDs are more effective than placebo for soft tissue injuries and osteoarthritis, although the improvement in osteoarthritis is small.
- The risk of adverse effects increases in patients over 70 years, and with duration and dose size.
- COX-2 NSAIDs have significantly less upper gastro-intestinal toxicity than non-selective NSAIDs.
- PPIs reduce the rate of upper gastrointestinal adverse effects. This is cost-effective for both non-selective and selective COX-2 NSAIDs in the UK, at £1000 per quality adjusted life year.
- The presence of dyspepsia does not predict peptic ulceration, bleeding, or perforation.
- Both classes of NSAIDs exacerbate ulcerative colitis and Crohn's disease and may injure the large intestine.

- Both NSAID classes, "with the apparent exception of Naproxen", are associated with an increased risk of myocardial infarction and coronary death. This translates into 3 more cardiovascular major events per 1000 patients taking COX-2 inhibitors or diclofenac for a year when compared to placebo. The risk is dose related.
- Recent meta-analysis shows no increased risk of stroke with either type of NSAID.
- The antiplatelet effect of low dose aspirin is blocked by non-selective NSAIDs except diclofenac (as long as it is taken 2 hours before the aspirin).
- COX-2 inhibitors do not block the antiplatelet effects of aspirin.
- Aspirin reduces FEV1 in 20% of adults and 5% of children with asthma and there is cross sensitivity with non-selective NSAIDs. Asthma "has not been produced" by COX-2 inhibitors.
- NSAIDs may cause abortion in the first trimester with an odds ratio of 2.43 compared to women not taking NSAIDs.
- Both classes of NSAIDs may delay labour and lead to premature closure of the ductus arteriosus.
- Both classes of NSAIDs can be used by breast feeding mothers. Just as well, because what with the ductus arteriosus being closed, you may not be able to get to the shops to buy any powdered milk.
- Paracetamol has an effect size of 0.14 compared with 0.29 for NSAIDs. NICE recommends combinations of paracetamol and opioids for osteoarthritis.
- Opioids are associated with high incidence of falls in the elderly. So easy does it with the opioids, Grandma!

QUIZ NO 7

Vertigo: Labyrinthitis is more common than vestibular neuritis.	☐ True ☐ False
Most patients with vertigo are likely to have BPPV or vestibular migraine.	☐ True ☐ False
Vestibular neuritis lasts for weeks and weeks and is likely to recur.	☐ True ☐ False
You can only get vestibular neuritis whilst standing in a vestibule.	☐ True ☐ False
The reason I keep making jokes about the Hallpike manoeuvre is because I don't have the first idea what it is.	☐ True ☐ False
Vertigo in BPPV is induced by head movements.	☐ True ☐ False
BPPV never recurs.	☐ True ☐ False
Patients with vestibular neuritis have nystagmus when looking forward.	☐ True ☐ False
IBS: IBS-C is the commonest subtype.	☐ True ☐ False
IBS is more common in people over 50.	☐ True ☐ False
Bowel transit times are normal in IBS.	☐ True ☐ False
In a recent fixture, the Rome III criteria were comprehensively outplayed by the Inter-Milan V criteria (2 goals in extra time)	☐ True ☐ False
It is probably a good idea to check FBC, CRP and coeliac antibodies in patients with suspected IBS.	☐ True ☐ False
The FODMAP diet may be effective and is very cost-	☐ True

effective for the NHS.	☐ False
Placebo response in IBS is about 40%.	☐ True ☐ False
Peppermint oil is jolly good - effective, harmless and cheap. Smells nice too.	☐ True ☐ False
It might be best to avoid those drugs from the USA.	☐ True ☐ False
Rifamixin is not related to TB or myxamatosis and does not make your urine go red.	☐ True ☐ False
Hypnotherapy is effective for IBS.	☐ True ☐ False
Acupuncture is good too, but you have to go to China to get it.	☐ True ☐ False
Necrotising fasciitis: Some patients may look absolutely fine but present with severe pain, and shriek alarmingly when you poke them with your finger.	☐ True ☐ False
Fever is likely.	☐ True ☐ False
Early prescription of antibiotics will prevent deterioration.	☐ True ☐ False
Checking ESR, CRP and sodium blood tests is advisable.	☐ True ☐ False
Mortality even with surgery is up to 40%, and delayed diagnosis increases mortality.	☐ True ☐ False
The hospital stay for necrotising fasciitis is at least 3 times the average, and then plastic surgery is needed. Alarming cost implications for your CCG.	☐ True ☐ False
Post traumatic stress disorder: Sufferers with PTSD remember every aspect of the precipitating event.	☐ True ☐ False
GPs are very good at making a diagnosis of PTSD.	☐ True

	☐ False
Don't ask patients about "re-experiencing" symptoms as this might upset them.	☐ True ☐ False
Patients with PTSD may present like patients with treatment-resistant anxiety or depression.	☐ True ☐ False
Peripheral arterial disease: Failure rate of surgical bypass grafts is no higher in those who continue to smoke.	☐ True ☐ False
Patients with diabetic foot ulcers will be seen immediately by a multidisciplinary diabetic foot team, and will simultaneously win the lottery without even buying a ticket.	☐ True ☐ False
Plaque rupture in a popliteal aneurysm can rapidly exacerbate symptoms.	☐ True ☐ False
Patients with critical limb ischaemia have a 5 year survival rate of about 50%.	☐ True ☐ False
Buerger's sign indicates peripheral arterial disease.	☐ True ☐ False
Buerger's sign indicates that you should pull over and buy yourself a Buerger.	☐ True ☐ False
ACE inhibitors reduce cardiovascular events in patients with PAD and have a number needed to harm (NNH) of 4, which means they're well worth a try.	☐ True ☐ False
Aspirin is still the best antiplatelet treatment in PAD.	☐ True ☐ False
Myasthenia gravis: Only 1% of patients with myasthenia gravis have a thymoma.	☐ True ☐ False
Myasthenia gravis can easily be recognised in the elderly, as the symptoms are so specific.	☐ True ☐ False
Muscle weakness is worse at the end of the day and may worsen with heat, infection, surgery, menstruation, etc. Just the same as if you haven't got myasthenia gravis.	☐ True ☐ False
Respiratory weakness may cause exertional dyspnoea.	☐ True

	☐ False
Bulbar weakness may cause your bulbs to go out.	☐ True ☐ False
The two main problems with the Ice test are firstly that most people don't like having their eyeballs frozen for three minutes, and secondly that you probably haven't got any crushed ice in your surgery.	☐ True ☐ False
Eventually, all myasthenia gravis patients will become seropositive.	☐ True ☐ False
The Tensilon test is still the gold-standard test for myasthenia gravis.	☐ True ☐ False
Thymectomy should always be performed if a thymoma is suspected, except under certain circumstances, when it shouldn't.	☐ True ☐ False
Quinine is useful for cramp in patients with myasthenia gravis.	☐ True ☐ False
Rotavirus vaccine: Rotavirus accounts for 80,000 GP consultations per year in the UK.	☐ True ☐ False
The sooner we get rid of all those consultations the better.	☐ True ☐ False
Rotavirus vaccination is over 90% effective.	☐ True ☐ False
Rotavirus vaccination is definitely not related to intussusception.	☐ True ☐ False
Cytisine: Cytisine is an alkuloid derived from the Ginko biloba tree.	☐ True ☐ False
It's made from canes cut from the tree - hence the famous film "Cytisine Cane".	☐ True ☐ False
Cytisine has been on the market in Eastern Europe since 1964, under the trade name Schmokoff.	☐ True ☐ False
Cytisine is as effective as other smoking cessation drugs	☐ True

and much cheaper.	☐ False
Thanks to its obvious advantages, and what with the Government being so anxious to save money on health care, cytisine should be licensed for smoking-cessation treatment in the UK any day now.	☐ True ☐ False
Autism: Children with autism rarely have other mental, behavioural or physical problems.	☐ True ☐ False
Secretin, chelation and hyperbaric oxygen therapies are well worth a try as they could help and have no known adverse effects.	☐ True ☐ False
Antipsychotics may be prescribed for challenging behaviour if psychological intervention cannot be delivered.	☐ True ☐ False
NSAIDs: Oral NSAIDs are more effective for knee osteoarthritis than quadriceps strengthening and walking.	☐ True ☐ False
NSAIDs only adversely affect the upper GI tract and not the large intestine.	☐ True ☐ False
Neither type of NSAID increases the risk of stroke.	☐ True ☐ False
Diclofenac and COX-2 inhibitors block the antiplatelet effects of aspirin.	☐ True ☐ False
Asthma may be worsened in 10% of adults by aspirin and non-selective NSAIDs.	☐ True ☐ False
COX-2 inhibitors could delay the start of the Oxford and Cambridge boat-race.	☐ True ☐ False

Management of nocturnal enuresis

BMJ 2013;347:f6259

Patrina H Y Caldwell, staff specialist paediatrician, Aniruddh V Deshpande, clinical and academic fellow in paediatric surgery and urology, Alexander Von Gontard, child and adolescent psychiatrist and paediatrician, chair and professor of child and adolescent psychiatry

- The prevalence of nocturnal enuresis is up to 20% in 5 year olds, 10% in 10 year olds and 2% in adults.
- It is defined as "intermittent involuntary voiding during sleep in the absence of physical disease in a child aged five years or more".
- It is more common in boys and associated with day-time urinary incontinence, faecal incontinence and constipation.
- It is associated with sleep disordered breathing and obesity.
- It occurs in 20-40% of children with psychological or behavioural disorders.
- Neuroimaging led by functional MRIs has shown that children with nocturnal enuresis have microstructural abnormalities and delay in maturation of the neuronal circuits in the prefrontal cortex.
- Nocturnal enuresis results from
  - Defective sleep arousal
  - Nocturnal polyuria
  - Bladder factors (eg. reduced bladder capacity or bladder overactivity)
- In primary care, take a detailed history of wetting, toileting patterns, fluid intake, co-morbidities, family situation. Examine for constipation, neurogenic or urological causes if there are lower urinary tract symptoms.
- Secondary enuresis occurs if bedwetting starts after attaining night dryness for at least six months.
- NICE advises "considering child maltreatment if secondary enuresis persists despite adequate assessment and management".

- In children with lower urinary tract symptoms in the daytime (non-monosymptomatic noctural enuresis) treatment should begin by dealing with the underlying daytime bladder problems.
- Nocturnal enuresis may be associated with low self-esteem or health related quality of life.
- Health seeking behaviour varies from country to country with 31.1% of parents consulting a health worker in a UK study.
- In children under 5, ensure adequate fluid intake, appropriate toileting behaviour, managing constipation and a trial of removal of nappies. No other treatment is advisable until 5 years of age.
- Refer if there are severe daytime symptoms, recurrent UTIs, abnormal rectal ultrasound, neurological problems, faecal incontinence, diabetes, behaviour, emotional or family problems or no response to treatment after 6 months.
- NICE advises conservative measures and, although evidence is that their efficacy is limited, they can be effective. They include support and education, avoiding caffeine-based drinks, adequate fluid intake, managing constipation and voiding every 2-3 hours in the day and avoiding "holding on".
- A small RCT found rewarding dry nights or lifting children during the night resulted in more children becoming dry.
- Alarm training is a first line treatment for nocturnal enuresis. The response is more gradual and sustained than for drugs with about two thirds of children becoming dry during treatment and about 50% remaining dry. Treatment should continue for a maximum of 16 weeks or until 14 consecutive dry nights.
- Desmopressin has been used to treat nocturnal enuresis for 40 years. Tablets have a lower risk of water intoxication than the nasal spray.
- Side effects of desmopressin include headaches, abdominal pain and emotional disturbances, but these are uncommon. Water intoxication may be minimised when children restrict drinking after taking desmopressin.
- A systematic review of 47 trials of desmopressin showed that less than half of those treated became dry and the relapse rate was high with no difference between desmopressin and placebo.
- A systematic review of 58 trials of imipramine showed that about a fifth of children became dry on treatment but the effect was not sustained after treatment was stopped.

- Due to its possible side effects, imipramine is only used for treating "resistant cases".
- A systematic review of acupuncture found that it was as effective as desmopressin.
- In one randomised controlled trial, hypnotherapy was as effective as imipramine with a lower relapse rate after cessation.
- Nocturnal enuresis has a spontaneous remission rate of about 15% a year.

Power to the people: what will bring about the patient centred revolution?

BMJ 2013;347:f6701

Paul Hodgkin, chief executive officer, Jeremy Taylor, chief executive officer

- The "patient centred revolution" is about patients acting as equal and informed partners in decision making, empowered by greater choice, information and the opportunity to give feedback.
- It is also about patients with long term conditions becoming managers of their own health.
- It's about holism, joined-up care and the importance of compassion, respect and dignity.
- "In other areas of life, society demands autonomy and responsibility of its citizens". In healthcare "illness can be inherently disempowering" and the paternalistic medical model persists.
- "Because the NHS is free at the point of use, behaviours and practices often seem to reinforce the dependency and passivity of the patient." This seems to imply that turning patients into "customers" is a way of empowering them, which certainly does seem to be a line of thought in healthcare policy.
- However, the growing population of people with multiple, long term conditions, disability and frailty requires "a primarily social not medical model". The keyword here is "integration", with different agencies working together to help patients (and their support networks) through the difficulties of ill health and old age (and taking the pressure off hospitals and care homes in the process).
- Transparency about performance "will continue to shift power from an often secretive and defensive NHS towards citizens". Oh will it? Has it shifted power in the education sector?
- "Digital technologies could change everything". Not if I keep dropping my mobile phone down the toilet, they won't.

- "Coming soon will be personalised genomics from companies like "23andMe," the ability for patients to upload data—such as blood pressure—on to their medical record, and social media supported behaviour change." Huh. So what?

- Let's say patients could upload their own weight readings onto their medical records. Your most overweight patient keeps coming to the surgery looking fatter and fatter, but according to his medical record he's getting thinner and thinner. In the end he explodes, but according to his record he's transformed himself into a subatomic wafer and slipped into another dimension. Where does that get you?

- "Medicalisation and marketisation could stymie moves to a more social and holistic model of care; transparency exposes inadequacy but does not itself generate the drive to overcome it. The digital revolution could empower or enslave the citizen. It is too early to tell." Amen to that.

Overdiagnosis: when good intentions meet vested interests—an essay by Iona Heath

BMJ 2013;347:f6361

Iona Heath, retired general practitioner

- Overdiagnosis and overtreatment have become embedded in the health care system.
- They have "permeated and polluted" the drug industry, medical research, payment systems, clinical practice and guidelines. They cause "waste and harm".
- The medical technology industry allows us to "measure and assign numbers to an ever increasing number of biometric parameters", which are usually normally distributed, "with one extreme representing a degree of abnormality that begins to correlate with symptoms and suffering". This extreme represents the cohort "that can be ameliorated or cured by medical treatment".
- There is a constant pressure to extend the range of "abnormal" into what was once considered to be normal.
- We have "erected an epidemic of disease without symptoms, defined only by aberrant biometrics".
- Labelling well people in this way may compromise their health and expose them to drugs with significant side effects.
- Jeremy Hunt instructed GPs to do more to prevent the health of older patients deteriorating. This distracts attention "from the government's failure to meet its own responsibilities for health protection through fiscal and legislative measures".
- Extending the range for what is considered abnormal expands markets for pharmaceutical interventions, thereby maximising profit.
- The Will Rogers phenomenon occurs when the range of a diagnostic category is extended and more "normal" people are included. Outcomes improve (basically because the new category-members weren't ill in the first place).

- Ethics and politics "are the only real defences that humanity has… to confront the abuse of power and money to the detriment of the weak and vulnerable".

- Medicine has colluded with society's search for "technical solutions to existential problems posed by the finitude of life and the inevitability of ageing, loss and death".

- "It is not knowledge that we lack. What is missing is the courage to understand what we know and draw conclusions" (Sven Lindqvist). (Basically, we have to accept our human lot.)

- Our greed for longevity may drive the "commercial imperatives of the pharmaceutical and medical technology industries".

- Countries in the OECD (us) "consume more than 80% of the world's health care resources but experience less than 10% of the world's disability adjusted years". This is untenable and unjust, especially if the rest of the world is pushed to or tries to follow.

- Overdiagnosis/labelling "deflects resources and attention from those most severely affected".

- Beware of Orwellian doublespeak as in 1984 – "War is peace, ignorance is strength, freedom is slavery…" Is health now disease?

- There are many measures held to be universally applicable in contemporary medicine, "whatever the circumstances of the individual life to which they are applied".

- "Our professional objectives are wider than the diagnosis and treatment of disease." The WHO definition of health suggests that we must acknowledge health is "about a person's potential for living which is a matter of autonomy and personal space" or having room to make choices.

- Our culture pays lip service to autonomy. In our culture, the state-approved "healthy choice" should be selected by all patients regardless of their circumstances.

- It's "good" to try and prevent disease, but the means are damaging and unidimensional, "and propagate an intensely normative view of what it means to be healthy and what human life should be".

- We need to develop a sense of "waste" in health care, morally and politically. We need to avoid harm.

- William James said "doubt and hope are sisters" – doubt existing explanations, and look for better ones that "could bring us hope".
- "Life must be understood backwards but lived forwards" (Kierkegaard).
- "I can't understand Kierkegaard either backwards or forwards" (Dr Hairy).
- For doctors, many of the links between cause and effect remain poorly understood. Much of what we now consider to be "true" will end up in Amitash Gosh's "graveyard of discredited speculations".
- "Health promotion… diminishes health and wastes resources" (James McCormick). GPs should "encourage people to lead lives of modified hedonism, so that they may enjoy, in full, the only life they are likely to have."
- To hell with the blood pressure. Pass me the crisps and cannabis.

How widespread is variant Creutzfeldt-Jakob disease?

BMJ 2013;347:f5994

Roland Salmon, retired consultant epidemiologist

- Variant Creutzfeldt-Jakob Disease (CJD) is the human form of bovine spongiform encephalopathy. It is one of several spongiform encephalopathies which result from a toxic build up of an aberrant form of a normal cellular protein, the prion protein.

- Basically, loads of cows (36,000) had it in 1992, but the human form is rare – 177 in the UK to date, and only one in the last two years.

- The "usual" sporadic CJD cases are found worldwide, with an incidence of 1/1,000,000 population/year. Sporadic CJD is transmissible by neurosurgery, by injection or by implantation of infected material, eg. human growth hormone from cadaveric pituitaries. (Yuk. Who'd agree to an implantation of that?)

- There is no convincing evidence of CJD transmission from surgery.

- Costly steps have been implemented to secure the blood (transfusion) supply and to reduce the risk from surgical instruments, but the risk/benefit of this approach depends on how many people in the UK are "infected".

- "Using archived tissue from appendicectomies and tonsillectomies", prion protein "was almost entirely absent from tonsils", but 1/4000 appendixes contained prior protein in one investigation, whereas Gill et al's survey of 30,000 appendixes showed a prevalence of 1/2000.

- In the UK patients with variant CJD have a modal death of 28 years and are mostly from northern England or Scotland. Confirmed cases have all been MM homozygous at the gene coding for prion protein.

- So… it is "possible" that the prion deposition in the appendix is of no relevance.

- People "deemed" to be at increased risk of CJD (CJD blood transfusion recipients or those who have come in contact with contaminated surgical instruments) could "possibly" transmit the disease.
- What is the phenotype and natural course of variant CJD in genotypes other than MM?
- What other animal prion diseases may be zoonotic?
- Prion proteins have been identified an amyloid-B in Alzheimers, Parkinson's and tau.
- How often, if ever, are any of the above transmissible?
- To hell with the CJD. Pass me a beef burger. And some more cannabis.

Novel drugs for treating angina

BMJ 2013;347:f4726

Daniel A Jones, research fellow, Adam Timmis, professor of clinical cardiology, Andrew Wragg, consultant cardiologist

- Initially advised antianginals include beta-blockers and calcium channel blockers, which reduce heart rate or vasodilate (respectively).
- After many years, NICE says that nitrates have insufficient evidence of efficacy (?) (are they too cheap?).
- NICE could not find enough evidence to make firm recommendations about choice of second line antianginals.
- If people with stable angina cannot tolerate beta-blockers or calcium antagonists then consider
  - Ivabradine
  - Nicorandil
  - Ranolazine
  - A long-acting nitrate (Baby Bio?)
- There is no evidence of benefit when three or more drugs are used.
- Nicorandil dilates coronary arteries.
- Ivabradine reduces heart rate.
- Ranolazine (and trimetazidine) have some, er, well, "metabolic modulation, increasing the efficiency of myocardial energy production".
- All of the above may have short term effects on exercise capacity and frequency of angina attacks (but not long term symptom relief or reduction in mortality).
- Ivabradine is ineffective in atrial fibrillation, but is useful in patients with contraindications to Beta-blockers or non-dihydropyridine calcium channel blockers. It is fine in COPD.
- Ivabradine reduces all cause mortality by 2% in patients with heart failure.

- No studies have been done of nicorandil monotherapy on fatal and non-fatal cardiovascular events.
- Ranolazine improves exercise performance the decreases frequency or angina with no prognostic benefits.
- Intolerance of ivabradine is similar to that of amlodipine – about 20%. Visual side effects are common with ivabradine.
- Nicorandil may cause oral ulcers, anal fissures and even fistulas – if so, stop it.
- Ranolazine may cause constipation, nausea and weakness.
- Trimetazidine may cause Parkinsonian symptoms and restless legs.
- Start low doses in patients over 75 years old of ivabradine and trimetazidine.
- "For patients taking ranolazine, an ECG should be obtained at baseline and follow up, to evaluate effects on QT interval".

Assessing risk of suicide or self harm in adults

BMJ 2013;347:f4572

Richard Morriss, professor of psychiatry and community mental health, Nav Kapur, professor of psychiatry and population health, Richard Byng, clinical senior lecturer

- Suicide is one of the top three causes of death in people aged 10-44 years in the world.
- Suicide rates have risen in the UK in the past few years, with the highest rates in 30-59 year old men. I'm not bloody surprised, either. Even higher rates for 30-59 year old doctors, I should think... Oops, that cannabis must be wearing off.
- Self harm is one of the five leading causes of hospital admissions and is associated with high risk of subsequent death.
- Current risk assessment tools are rubbish.
- GPs should follow "a structured pattern", as described below. Briefly:
    - Clarify current problems
    - Assess mental illness
    - Assess current thoughts, plans and intent
    - Consider other risk factors
    - Summarise and agree action plan
- "The assessment of risk must be individualised and take into account the patient's mental state and social context". Check for depression, delusions, hallucinations and substance misuse.
- Ask in sequence, according to response:
    - How do you see the future?
    - Have you been feeling that life is not worth living?
    - Have you thought about ending your life?
    - Have you made plans to end your life? Tell me more about your plans...
    - Would you carry out these plans? What would make this more or less likely?

- Consider life plans, support of others, previous attempts, if the patient is under 25 and on antidepressants, access to means of suicide, notes, will-changes, dependent children...
- After self-harm consider
  - Lethality of attempt
  - Expectations of outcome
  - Precautions of discovery
  - Mental state at time of self harm (mood, drugs, alcohol)
- The balance of risk and protective factors will vary from person to person. Two people the same age and social background may have radically different perspectives concerning the same life situation.
- If a person refuses help if he/she is at high risk of suicide, seek urgent help from mental health services if there are "reasonable grounds to suspect a mental disorder" (including personality disorder), and consider involuntary detention.
- If the person's mental state is related to drugs or alcohol alone, then involuntary detention cannot be used in England. If you live close to the Scottish border, you could whip him across to the other side and get him locked up there.
- Assessment of suicide risk is not precise, and risk factors may change rapidly over short periods.

Investigating stable chest pain of suspected cardiac origin

BMJ 2013;347:f3940

Declan P O'Regan, consultant radiologist, Stephen P Harden, consultant adiologist, Stuart A Cook, professor of cardiology

- NICE advice in 2010 no longer recommended exercise ECG to investigate stable chest pain in people without known coronary artery disease.

- Most patients will have symptomatic coronary artery disease diagnosed by non-invasive cardiac imaging with invasive angiography reserved for higher risk patients being considered for revascularisation.

- CT coronary artery calcium score is a screening test for coronary artery disease (98% sensitivity, 40% specificity). Calcium scoring has a negative predictive value of 99% and a normal study result predicts a 2% cardiac event rate over 5 years.

- Calcium scoring radiation dose is low. It will rule out symptomatic coronary artery disease in at least 70% of those referred.

- CT coronary angiogram is sensitive (98%) and specific (92%) and is used to see if people with a CT calcium score of 1-400 have obstructive coronary disease. IV contrast is used. Patients often need heart rate control with beta-blockers or vasodilation with GTN.

- Functional cardiac imaging is used to investigate patients with a 30-60% probability of coronary artery disease. Functional imaging includes
  - Perfusion imaging
  - Wall motion imaging

- Perfusion imaging uses an intravenous vasodilator which increases the difference between perfusion of normal myocardium and that supplied by a stenosed artery.

- Myocardial perfusion scintigraphy and cardiac magnetic resonance imaging have very similar sensitivities (about 90%)

and specificity (about 78%). Adenosine is contraindicated in severe asthma and 2nd and 3rd degree heart block. Caffeine should be avoided for 24 hours before adenosine due to competitive inhibition.

- Dobutamine can be used to induce wall motion abnormality within ischaemic myocardium. Abnormalities may be seen on echocardiography or cardiac magnetic resonance. Dobutamine is contraindicated in severe aortic stenosis and obstructive hypertrophic cardiomyopathy. Sustained ventricular tachycardia occurs in 1.7/1000 patients.
- If the estimated risk of coronary artery disease is 61-90%, patients should be considered for coronary angiography and revascularisation. This is the gold standard investigation (still). It may be difficult to define the importance of intermediate stenosis on angiogram, so the flow limiting effect can be assessed by a pressure catheter. The overall complication rate is 7.4/1000 with a mortality of 0.7/1000.

Gout

BMJ 2013;347:f5648

Edward Roddy, senior lecturer in rheumatology, Christian D Mallen, professor of general practice research, Michael Doherty, professor of rheumatology

- If gout is undertreated, continued crystal deposition "can" cause irreversible joint damage.
- About (or "only") 10% of people with hyperuricaemia develop gout.
- If it involves the first metatarso-phalangeal joint it can be called "podagra" and occurs in up to 78% of first attacks.
- Attacks of gout reach "peak intensity within 12-24 hours" and resolve in one-two weeks.
- If untreated, a second attack occurs within 2 years. Recurrent attacks "may become more frequent" and chronic tophaceous gout may occur.
- 70% of uric acid derives from breakdown of endogenous purines and 30% comes from dietary purines. 70% of uric acid is excreted through the kidney.
- Metabolic syndrome is strongly associated with gout.
- The risk of gout is "related" to alcohol consumption, especially beer (but not wine). The risk is also increased with red meat and "seafood". Dairy products are productive. Sugar sweetened drinks, especially those containing fructose, increase the risk (second only to beer). Coffee is protective.
- There is a genetic link, some rare enzyme defects and some affective renal clearance/transport.
- Some studies "found a trend" towards a higher risk of gout in patients taking loop and thiazide diuretics. There are no RCTs to test the influence of stopping diuretics in people with gout – consider it.
- The effect of low dose aspirin is "thought" to be insignificant and should be continued for cardiovascular prophylaxis.

- Comorbidities are associated with gout (hypertension, CKD, obesity, diabetes, congestive heart failure and ischaemic heart disease). These may limit treatment options.
- A clinical diagnosis is usually accurate and "podagra" has high sensitivity and specificity. Joint aspiration may be useful if there is clinical doubt and also excludes pseudogout or septic arthritis.
- There is no evidence to support drug treatment of people with asymptomatic hyperuricaemia and people with gout "may have normal serum urate concentrations", especially in an acute attack.
- First line oral drugs are NSAIDs or colchicine. No NSAID is better than any other.
- Systematic reviews and RCT suggests indomethacin was comparable to etoricoxib and that indomethacin and naproxen are "as effective as oral prednisolone". Indomethacin "is best avoided"(!).
- Colchicine inhibits inflammatory cell mediated responses to phagocytosis of urate crystals. Lower doses of colchicine are as effective and more tolerable than normal high dose regimes.
- The low dose regime is 1.2mg initially, followed by 600mcg after one hour.
- Several drugs increase the risk of colchicine toxicity.
- IM or oral or inta-articular injection of steroids are all effective and 2 RCTs suggest 30mg of prednisolone for 5 days is as effective as NSAIDs.
- Application of ice to the affected joint is effective.
- Offer urate lowering drugs to people with recurrent actue gout, tophi, x-ray damage, renal insufficiency or uric acid urolithiasis.
- The "threshold at which recurrent of acute attacks warrants treatment is controversial".
- "Most people wish to receive" urate lowering therapy "when fully informed".
- Allopurinol should be increased monthly until serum urate is below 360 micromols/L. "Many patients need 400-500mg of allopurinol daily to (adequately) reduce uric acid."
- Lower doses are recommended in people with renal failure, as there is a risk of hypersensitivity syndrome (skin reactions, liver and kidney problems, life threatening).

- When allopurinol is prescribed acute gout attacks may be prevented by an NSAID or low dose colchicine for up to 6 months.
- Febaxostat is approved by NICE for people who cannot tolerate allopurinol or for whom allopurinol is contraindicated.
- There is no evidence that wrapping your foot in a huge white bandage, propping it on a low stool, and going "Oooh-hoo-hoo-hoo!" every time Oliver Hardy trips over it is at all helpful in the treatment of gout. On the other hand, it certainly is jolly funny.
- To hell with the gout. Pass me a big glass of port.

Long acting β2 agonists in adult asthma

BMJ 2013;347:f4662

Graeme P Currie, consultant chest physician, Iain Small, general practitioner, Graham Douglas, consultant physician

- Long acting beta2 agonists (LABAs) have a bronchodilator and "airway stabilising" effect.
- The effect of salmeterol and formoterol lasts for about 12 hours. Salmeterol works within 10-30 minutes and formoterol in 1-3 minutes.
- Large RCTs and systematic reviews show that LABAs, added to adults already using inhaled steroids, reduce symptomatic use of relief inhaler and exacerbations and improve quality of life and lung function.
- Because of the rapid onset of action of formoterol, some inhalers containing both formoterol and inhaled steroids can be used "as required" as well as regularly ("maintenance and reliever therapy"). RCTs have suggested that maintenance reliever therapy significantly reduces exacerbations requiring oral steroids. Symbicort 100/6 and Symbicort 200/6 is licensed for this purpose, and so is Fostair. Not Ryan Air, though.
- Side effects of LABAs include tachycardia, tremor, muscle cramps, hypokalaemia and paradoxical bronchospasm and "nervousness".
- Some studies have shown LABAs may result in deteriorating asthma control, possibly due to B2 adrenoreceptor "down regulation" and less stringent use of inhaled steroids.
- An FDA meta-analysis showed that patients using LABAs "experienced significantly more 'events'" including death, intubation and hospital admission. None of the "events" occurred in patients using LABAs and inhaled steroids in a combination device.
- It is "highly likely that worsening asthma control is related to under-treatment with inhaled corticosteroids rather than a direct effect of LABA use". MHRA advises that LABAs, when used with inhaled steroids, outweigh any apparent risks. Combined

inhalers are probably the best was to ensure inhaled steroids are not reduced by patients starting LABAs.

- The inhaled corticosteroid dose at which LABA should be added is not established, but a meta-analysis shows the dose response curve plateaus at 100-200mcg of fluticasone daily, and little extra benefit is observed beyond 500mcg a day (equivalent to 1000mcg of betamethasone).

- After starting LABAs consider review six weeks laters and using the validated asthma control test in primary care.

- LABAs are the preferred add-on treatment for adults receiving inhaled steroids, compared with Leukotrine receptor agonists (LTRA), "although the greater benefit is modest" (about a 2% difference in exacerbations requiring steroids).

- However, LTRAs exhibit "significantly superior" effects in terms of attenuating airway hyper-responsiveness and reducing markers of underlying inflammation.

Diagnosis and management of varicose veins in the legs: summary of NICE guidance

BMJ 2013;347:f4279

- Grace Marsden, senior health economist, Mark Perry, research fellow, Kate Kelley, associate director, Alun H Davies, professor of vascular surgery and honorary consultant surgeon on behalf of the Guideline Development Group
- Those of you who are paying very close attention may be thinking to yourselves "Wait a minute! Dr Hairy already summarised a BMJ article about varicose veins! Don't tell me he's just going to start recirculating the same summaries over and over again, the lazy old geezer! Is this what I paid my £40 for? I want my money back! Etc etc..."
- However, that was a very different article. Very very different. Yes, it was about varicose veins, but it was about them in a completely different way. No resemblance, in fact. And, er, you can't have the money back because I've already spent it.
- Now read on...
- About a third of the UK population have varicose veins.
- About 10% of people with varicose veins develop skin changes, while 3% may develop ulcers.
- There is substantial variation across the UK as to who qualifies for referral and how varicose veins are treated.
- "Refer people with bleeding varicose veins to a vascular service immediately."
- Refer people with varicose veins that are causing pain*, heaviness, swelling or itching or skin changes (pigmentation of eczema).
- *Pain isn't mentioned as a criterion for referral in our local referral guidelines.
- Refer patients with superficial vein thrombosis "and suspected vein incompetence".
- Refer patients with a venous ulcer.

- Endothermal ablation (radiofrequency or laser) may be offered.
- If endothermal ablation is unsuitable, ultrasound guided foam scleropathy may be offered, and if this is unsuitable, truncal vein stripping and treatment of incompetent tributaries may be offered.
- Compression bandaging/hosiery should not be used for more than seven days after treatment.
- Patients must know that new varicose veins may develop after treatment, and more than one session of treatment may be needed.
- The chance of recurrence after treatment for recurrent varicose veins is higher than for primary varicose veins.
- Varicose veins occurring in pregnancy may regress postnatally. So, to hell with the varicose veins: I'm off to get myself pregnant.

QUIZ NO 8

Nocturnal enuresis: Nocturnal enuresis never occurs in adults.	☐ True ☐ False
In non-monosymptomatic nocturnal enuresis, always start treatment for the underlying bladder problem.	☐ True ☐ False
There is no need for GPs to arrange renal ultrasound scans for children with nocturnal enuresis.	☐ True ☐ False
Refer if no response to treatment after 3 months.	☐ True ☐ False
It is sensible to advise fluid restriction and to "hold on" for as long as possible.	☐ True ☐ False
Desmopressin has a high relapse rate.	☐ True ☐ False
Imipramine has loads of side effects and a high relapse rate.	☐ True ☐ False
Acupuncture and hypnotherapy are as effective as desmopressin or imipramine, with a lower relapse rate.	☐ True ☐ False
Alarm training consists of sneaking into the child's bedroom at night and shouting "Boo!"	☐ True ☐ False
Patient centred revolution: Is about informed, shared decision-making, empowering patients and making them managers of their own health.	☐ True ☐ False
It's disempowering for patients to be given their health care free. Customers are able to choose what they buy and where they buy it. Therefore if patients were more like customers, they would have more empowerment and freedom.	☐ True ☐ False
The future of health is primarily based on a social model.	☐ True ☐ False
The future of health is primarily based on loads of apps	☐ True

and digital technology and stuff like that.	☐ False
Whatever we think the future of health care is going to be like, when we actually get there it's probably not going to be like that.	☐ True ☐ False
Overdiagnosis: "Preventive health care" and "an epidemic of disease without symptoms" are different ways of describing the same thing.	☐ True ☐ False
The Will Rogers phenomenon occurs when disease categories are expanded, well people are told they are ill, and outcomes in disease categories appear to improve.	☐ True ☐ False
The Roy Rogers phenomenon occurs when men wear ten-gallon hats, ride their horses to the tops of dusty hills, and shout "Yee-ha!" This can also lead to improved outcomes.	☐ True ☐ False
It's a good thing to doubt existing explanations and look for better ones.	☐ True ☐ False
GPs should encourage people to lead hedonistic lives, enjoy themselves to the full and stay away from the surgery at all costs.	☐ True ☐ False
Creutzfeldt-Jakob disease: Spongiform encephalopathies result from a toxic build up of prion proteins.	☐ True ☐ False
The usual sporadic CJD cases have an incidence of 1/1,000,000 per year, so you are more likely to get CJD than win the lottery.	☐ True ☐ False
In the UK, patients with variant CJD are mainly from the southern counties and/or have a genetic predisposition.	☐ True ☐ False
Drugs for angina: Adding a third drug will be of enormous benefit to patients already taking two antianginal drugs.	☐ True ☐ False
Nicorandil dilates coronary arteries, ivabradine reduces heart rate, and ranolazine works by magic.	☐ True ☐ False
All of the above drugs make your patient with angina feel better and live longer.	☐ True ☐ False
If you develop mouth ulcers or anal fissures when taking micorandil, just keep taking it because the effects will	☐ True

eventually wear off.	☐ False
Suicide and self harm: 30-59 year old men have the highest suicide rates in the UK. That's because their wives and children take all the money, and then won't even let us watch what we like on the telly. Are you with me, brothers?	☐ True ☐ False
Self harm is not associated with a high risk of subsequent death.	☐ True ☐ False
People are likely to have the same perspective concerning the same life situation, with no variation between individuals.	☐ True ☐ False
If there is high risk of suicide and the patient refuses help, consideration should be given to involuntary detention, even if their mental state is related to drugs or alcohol.	☐ True ☐ False
Investigating chest pain: CT coronary artery score has a good negative predictive value.	☐ True ☐ False
IV contrast is used in CT coronary angiogram, and beta-blockers or GTN may be needed.	☐ True ☐ False
IV vasodilators are used in perfusion imaging.	☐ True ☐ False
It is best to avoid strong coffee before IV adenosine, but a good stiff whisky may improve the vasodilating effect.	☐ True ☐ False
Wall motion imaging using dobutamine is completely safe and never causes sustained ventricular tachycardia.	☐ True ☐ False
You are more likely to die having coronary angiography than to be killed by lightning strike.	☐ True ☐ False
Gout: 70% of uric acid comes from dietary prunes.	☐ True ☐ False
70% of uric acid comes from dietary purines.	☐ True ☐ False
If you accidentally drink 3 pints of beer whilst eating a prawn cocktail and a bacon sandwich, you might be able to stave off an attack of gout by ordering a cheese platter, a	☐ True ☐ False

glass of milk and a double espresso.	
Podagra should not be regarded as an alternative to viagra, as it will make a gentleman limp.	☐ True ☐ False
NSAIDs and colchicine are first line drugs for acute gout, but prednisolone is much less effective than NSAIDs.	☐ True ☐ False
Allopurinol should be offered forcefully to anyone who has had 2-3 acute attacks of gout.	☐ True ☐ False
The singular of tophi is toffee.	☐ True ☐ False
Long acting Beta2 agonists: Maintenance reliever therapy reduces exacerbations requiring oral steroids, and the inhalers licensed for this are symbicort and fostair.	☐ True ☐ False
Side effects of LABAs include hypokalaemia and paradoxical bronchospasm.	☐ True ☐ False
Side effects of LLAMAs include spit all over your jumper and fluff on your trousers.	☐ True ☐ False
Other side effects of LABAs include worsening asthma control and, er, a cessation of life.	☐ True ☐ False
LABAs are better add on treatment than leukotrine receptor agonists (LTRAs).	☐ True ☐ False
LTRAs are better add on treatment than LABAs.	☐ True ☐ False
LLAMAs are better than either of them.	☐ True ☐ False
Varicose Veins: Refer people with varicose veins if they are heavy, swollen, itchy, pigmented, eczematous, bleeding or ulcerated.	☐ True ☐ False
Also refer them if their varicose veins are painful. Just don't tell the CCG.	☐ True ☐ False
Endothermal ablation may be offered if available, or ultrasound sclerotherapy, or surgical stripping, although it	☐ True ☐ False

really depends on where you live.	
After treatment, compression hosiery should be worn for at least 6 weeks.	☐ True ☐ False
Varicose veins occuring in pregnancy may regress postnatally. But the other effects of parenthood - such as poverty, baldness and dementia - may never wear off.	☐ True ☐ False

The role of dipeptidyl peptidase-4 inhibitors

BMJ 2012;344:e1213

Daniel Lasserson, senior clinical researcher, Jonathan Mant, professor of primary care research

- DPP-4 inhibitors reduce the breakdown of the glucose responsive incretin hormones, mainly gulcagon-like petide-1 (GLP-1).
- They "normalise glucose homeostasis without increasing body weight".
- They may be less likely to cause hypoglycaemia than other oral hypoglycaemic agents.
- They are associated with favourable changes in cardiovascular risk factors.
- DPP-4 inhibitors are less effective than metformin as monotherapy.
- Combined with metformin, DPP-4 inhibitors are less effective than metformin combined with other hypoglycaemic agents at lowering HbA1c.
- NICE recommends a sulfonylurea as second line agent after metformin and insulin as third line, with DPP-4 as third line if insulin is unacceptable.
- NICE recommends DPP-4 as second line treatment if the patient is at high risk of hypoglycaemia or cannot tolerate a sulfonylurea.
- There is "no compelling evidence that DPP-4 inhibitors play a major role in the management of type 2 diabetes" but the low incidence of hypoglycaemia may be useful in some patients.
- The risk of hypoglycaemia is important when deciding which hypoglycaemic drug to prescribe, but "there is no standard definition of hypoglycaemia in diabetes trials".
- Early and sustained HbA1c reductions are associated with long term cardiovascular risk reduction. Short term HbA1c reduction "is not a reliable indicator for reduction of cardiovascular risk".

Investigating the pregnant woman exposed to a child with a rash

BMJ 2012;344:e1790

Eithne MacMahon, consultant and honorary senior lecturer

- A pregnant woman comes to see you in a state of high anxiety, because her sister came to visit yesterday, bringing with her a feverish three-year-old daughter, who on closer inspection turned out to have a horrible-looking rash all over her body. What do you do?
- The basic outline goes like this:
    - You have to try to establish what the child was suffering from;
    - You have to look at the pregnant patient's immunisation-and-illness history, to find out if she's already be immune to whatever the child had;
    - If she turns out not to be immune, you may have to offer her immunisation.
- Rashes in children that put pregnant women at risk include those due to measles, rubella, parvovirus B19 and varicella zoster.
- 50% of pregnant women remain susceptible to parvovirus B19.
- "Exposure" to varicella means face to fact contact, or 15 minutes in the same room. The same criteria are used for rubella and parvovirus. Household exposure causes high transmission risk.
- Where feasible secure a laboratory diagnosis for the child.
- If the child's rash was maculopapular, measles is the most urgent consideration as immunoglobulin must be administered to the pregnant woman within 6 days.
- Measles and rubella in the index case (child) can be diagnosed by NAT testing (nucleic acid amplification technology) of saliva or throat swab.
- Parvovirus IgM is usually detectable at the onset of rash in symptomatic cases, but a positive result requires confirmation.

- NAT testing can detect varicella zoster virus DNA in crusted lesions "deemed to be no longer infectious". Older tests are less specific or sensitive.
- For measles or rubella, previous MMR vaccination or previous positive IgG test result provides "satisfactory evidence of immunity".
- For parvovirus B19, serological testing is essential unless another diagnosis has been confirmed in the child.
- If a pregnant woman's measles immunity is unknown, check measles IgG. If this is negative, HNIG (human normal immunoglobulin) may prevent or attenuate maternal measles, if given within 6 days of exposure.
- If immunity to rubella is in doubt, the woman should be tested for rubella IgG and IgM. If the rubella IgG level is below 10 IU/mL then the mother may be susceptible to rubella. If the IgG and 1gM are negative, repeat the test at least a month after exposure – if IgM is detected, "seek specialist advice".
- If the woman may have been exposed to parvovirus B19, test her for parvovirus IgG and IgM. If IgG is detected and IgM not detected within a month of exposure, she can be reassured. In cases of maternal parvovirus B19 infection, close monitoring for hydrops fetalis and the need for intrauterine transfusion is required.
- After exposure to a vesicular rash, assess the woman's varicella zoster virus status, as she would benefit from VZIG (varicella zoster immunoglobulin) if she is susceptible.
- A history of 2 varicella vaccines or a history of chickenpox or shingles (in temperate countries) is likely to signify immunity to chicken pox.
- If there is no history to suggest immunity, test for varicella zoster IgG.
- VZIG may prevent or ameliorate severe maternal chickenpox or maternal death. (Anything that can ameliorate death is doing pretty well.) It is not clear that VZIG can prevent congenital chickenpox. VZIG can be given within 10 days of the onset of rash in the index case (in household contacts).
- Whether VZIG is given or not, the woman should seek medical advice if she develops an illness with a rash.

- Testing can be expedited if antenatal bloods are routinely stored in the laboratory.

Diagnosis and management of headaches in young people and adults: summary of NICE guidance

BMJ 2012;345:e5765

Serena Carville, senior research fellow and project manager, Smita Padhi, research fellow, Tim Reason, statistician and health economist, Martin Underwood, professor of primary care research on behalf of the Guideline Development Group

- Infection, tumour, bleeding and arteritis are the serious causes of headaches. (Some of them are more serious than others, obviously.)
- The major health/social burden is caused by cluster headaches, migraine, tension headaches and over-use of medication.
- Further investigation or referral of patients with headaches should be considered if there is
    - fever
    - sudden onset (maximum intensity by 5 minutes)
    - neurological or cognitive deficit
    - changed consciousness level
    - change in personality
    - head trauma in last 3 months
    - headache triggered by exercise, cough, valsava, sneezing or change in posture
    - symptoms of giant cell arteritis or acute glaucoma
    - change in characteristics of headache
- New onset headaches may also merit further investigation or referral if there is compromised immunity, previous history of malignancy (especially under 20 years old) and/or vomiting without other obvious cause.
- Overuse of triptans, opioids, ergots, paracetamol, aspirin and NSAIDs may cause headaches (if used 10-15 days a month or more).

- In tension-type headaches consider NSAID or paracetamol, but avoid opioids for acute treatment. Consider acupuncture.
- In migraine offer oral triptan AND an NSAID or oral triptan AND paracetamol for acute attacks.
- For 12-17 year olds consider nasal triptan rather than oral triptan.
- Offer topiramate or propranolol for prophylactic treatment.
- Topiramate is associated with foetal abnormalities and can impair effectiveness of hormonal contraceptives.
- Acupuncture over 5-8 weeks or gabapentin (up to 1200mg/day) may be considered if topiramate or propranolol are unsuitable or ineffective.
- Riboflavin 400mg once a day "can reduce the frequency and intensity of migraine in some people".
- In those who have migraine with aura, do not routinely offer combined hormonal contraceptives.
- For predictable menstrual related migraine unresponsive to standard acute treatment, consider frovatriptan 2.5mg b.d. or zolmitriptan (up to 2.5mg tds) on the days migraine is expected.
- In cluster headaches offer oxygen and a subcutaneous or nasal triptan for acute treatment. Verapamil can be considered for prophylactic treatment (seek advice on ECG monitoring).
- In medication overuse headache, advise people to stop taking all overused headache drugs immediately (who'd have thought it, eh?) and for at least a month, but warn them about worsening headaches initially and possible withdrawal effects. Resolution of headaches occurs in only half of those who succeed in stopping overused medication.

Telemonitoring for patients with COPD

BMJ 2013;347:f5932

Rachel Jordan, senior lecturer, Peymane Adab, professor, Kate Jolly, professor

- Cochrane reviews conclude that education on COPD self management "is associated with" a reduced risk of hospital admission.
- Pinnock and colleagues' trial in Lothian randomised patients into telemonitoring or conventional self monitoring.
- The telemonitoring group did daily touch screen questionnaires and used "an instrument to measure oxygen saturation" (a set of bagpipes?).
- 85% of patients randomised to telemonitoring received the equipment and completed the training.
- "After 12 months, no difference was seen in hospital admissions for COPD between the two groups".
- No differences were seen in health related quality of life, anxiety or depression, self efficacy, knowledge or adherence to drugs.
- There were large numbers of contacts through the alert system (24/person/year) and many of these were "false alarms". There were also more contacts not related to alerts in the telemonitoring group.
- "Counterintuitively, patients with milder disease and those with higher depression scores seemed to have higher admission rates with telemonitoring".
- The trial suggests that telemonitoring of COPD patients "is costly and ineffective... creates a high workload, is expensive, and may result in a large number of false positive alerts and overtreatment. "
- In your face, Department of Health!

India has a problem with palm oil

BMJ 2013;347:f6065

**Bhavani Shankar, professor of food, agriculture, and health,
Corinna Hawkes, head of policy and public affairs**

- Basu and colleagues simulated the effects of a 20% tax on palm oil on serum cholesterol and mortality from coronary heart disease and cerebrovascular disease in India.
- Palm oil is under-studied, and is the world's most consumed oil, of great importance to the health, environment and economies of Asia.
- The authors estimate that palm oil tax would have a modest impact on hyperlipidaemia and mortality, but their estimate looks at household consumption and not the use of palm oil in food processing.
- Palm oil is "good" for processed foods as foods processed with saturated fats have longer shelf lives.
- Taxation of palm oil might prompt manufacturers to "simply switch to another source of unhealthy fats - trans fats".
- Contrary to what you might think, trans fats are not chubby blokes dressed up as women. Those are fat trans.
- Trans fats are present in "vanaspati", a vegetable ghee used in Asian households. Vanaspati contains up to 50% trans fat content. "Whereas vanaspati typically consisted of other oils in the past, palm oil and its fractions are now reported to form a major proportion of oils used in its manufacture."
- If a tax on palm oil reduced its presence in vanaspati, further action might be required to prevent other equally-harmful oils from taking its place.
- Palm oil has become so financially successful because its production costs are low.
- 2.7 – 4 million hectares of forest were lost to palm oil production in Indonesia and Malaysia between 1990 and 2005.
- Increasing tax on palm oil might reduce palm oil expansion elsewhere in the world and thus benefit the environment.

- Is it wrong of me to fancy a curry after writing this?

Severe hypotension associated with α blocker tamsulosin

BMJ 2013;347:f6492

Jorge Ramirez, professor of pharmacology

- The prevalence of lower urinary tract symptoms varies from 15% - 60% of men over 40 years old.
- Bird and colleagues found a significant association between starting or restarting tamsulosin and severe hypotension (severe enough to require admission).
- "Tamsulosin was introduced in 1996 and marketed as a major innovation among α blockers because it was associated with a lower frequency of orthostatic hypotension than other drugs in this class."
- Tamsulosin "dominates the global market for the treatment of benign prostatic hyperplasia (BPH)". It is available over the counter in the UK.
- Apparently tamsulosin levels concentrate in the prostate and it binds preferentially to prostatic alpha receptors.
- Systematic reviews suggest that tamsulosin is "moderately effective" for LUTs.
- Tamsulosin does not improve clinical outcomes of BPH.
- The hypotensive effect of alpha-blockers may be exacerbated by co-treatment with PDE-5 inhibitors.
- Tadalafil has recently been approved for treatment of BPH and fixed combinations of alpha-blockers and tadalafil are being developed.
- The safety evidence for combined tamsulosin and PDE-5 inhibitors is based on small trials with little follow-up and some registered studies of this combination have never been published.
- Conclusion: drugs which are marketed as "major innovations" often turn out not to be quite as great as the marketing would have us believe.

Secondary prevention for patients after a myocardial infarction: summary of updated NICE guidance

BMJ 2013;347:f6544

Katie Jones, project manager, Leanne Saxon, research fellow, William Cunningham, retired general practitioner, Phil Adams, emeritus consultant cardiologist on behalf of the Guideline Development Group

- Shorter hospital stays following myocardial infarction (MI) mean that GPs have a greater responsibility for promoting cardiac rehabilitation and prescription and monitoring of ongoing drug therapy.
- "Cardiac rehabilitation should be offered to all people who have had an MI".
- Patients should be told that cardiac rehab reduces risk of death and having another MI, and also improves quality of life.
- Rehab should start "within 10 days" of discharge.
- Patients should be advised to eat a Mediterranean diet, lose weight, stop smoking, drink no more than 14-21 units of alcohol per week and do 20-30 minutes of exercise a day. In other words, the same stuff you told them before they had the MI.
- Do not advise omega 3 fatty acid capsules and supplemental foods.
- Titrate ace inhibitors upwards at short intervals and ensure complete titration in 4-6 weeks after discharge.
- Offer angiotensin receptor blockers to people intolerant of ACE-inhibitors.
- Start an aldosterone antagonist 3-14 days post MI for patients with "symptoms or signs of heart failure and left ventricular dysfunction".
- Offer low dose aspirin after an MI and continue it long-term.
- Consider clopidogrel if aspirin is contraindicated

- Offer 12 months' clopidogrel combined with aspirin in patients who have had ST elevation MI and received a stent, or non-ST elevation MI.
- Tricagrelor (12 months) plus aspirin should be given to those with ST elevation MI awaiting PPCI (primary percutaneous coronary intervention) and those with NSTEMI.
- Offer beta blocker as soon as possible after an MI. Well, maybe not before the patient gets into the ambulance.
- Titrate beta blockers to the maximum tolerated dose.
- Continue beta blocker for at least 12 months after an MI in people without left ventricular systolic dysfunction or heart failure.
- Continue beta blocker indefinitely in people with left ventricular systolic dysfunction.

Idiopathic pulmonary fibrosis

BMJ 2013;347:f6579

Owen J Dempsey, consultant chest physician, David Miller, clinical lecturer

- Idiopathic pulmonary fibrosis is an interstitial lung disease of unknown cause, with a median survival of 3 years after diagnosis.
- There are about 5000 new cases per year in the UK.
- Median age of onset is 70 years, and two thirds of those diagnosed are smokers.
- There is a low awareness of the disease among health professionals.
- It is uncommon under 45 years old, and the key clinical finding is "velcro"-like crackles audible over lower lobes and the axillae.
- Finger clubbing is found in 40% of patients.
- Diagnostic delay is common.
- Chest x-ray, high resolution CT imaging and, if in doubt, surgical biopsy confirms diagnosis.
- Pulmonary fibrosis may also be caused by connective tissue diseases, exposure to organic and inorganic dusts, and drugs such as nitrofurantoin and amiodarone.
- Lung transplantation may be recommended in selected patients. Pulmonary rehabilitation, oxygen therapy and palliative care may help.
- Interferon gamma, bosentan, macitentan, ambrisentan, sildenafil, imatinib, warfarin, oral corticosteroids and immunosuppressants are all ineffective.
- N-acetyl cysteine significantly reduces the rate of decline of lung function, but with no effect on mortality.
- Nintedanib showed a reduction in decline in lung function, fewer exacerbations, and preserved quality of life in one study.
- Co-trimoxazole had no effect on lung function but improved quality of life and reduced mortality in those adhering to treatment in a recent UK study.

- Pirfenidone has recently been approved by NICE. It has antifibrotic, anti-inflammatory and antioxidant properties, "although its precise mechanism of action remains elusive".
- A Cochrane review showed pirfenidone reduced the risk of disease progression by 30% but had no effect on mortality. It is recommended for mild to moderate disease. Initial estimated cost of pirfenidone was £26,000 per year.

An adult with a neck lump

BMJ 2013;347:f5473

Raghav C Dwivedi, senior head and neck fellow, Liam Masterson, specialty registrar in ear, nose, and throat surgery, Mostayn Alam, GP registrar, Piyush Jani, consultant head and neck surgeon

- Congenital neck lumps can present in adults and often enlarge over a long time period.
- Ask about foreign travel, insect bites, skin lesions, head, neck or scalp inflammation or infection, immunosuppression and previous cancers.
- Posterior triangle lumps mainly result from local inflammation (and also infection mononucleosis or lymphoma).
- Supraclavicular lumps may be metastases from lung, breast or abdominal cancer.
- Look in the mouth, especially if the lump is firm or fixed or rapidly growing.
- Purplish nodules can be associated with Kaposi's sarcoma or, when combined with ulcerative skin lesions overlying a lymph node, may suggest Mycobacterium avium-intracellulare infection.
- Red flag symptoms include dysphagia or pain on swallowing, voice change, persistent sore throat, unilateral nasal obstruction, epistaxis, otalgia or glue-ear, weight loss, fever, night sweats and bone pain. Consider urgent referral.
- Any new neck lump that persists beyond 4-6 weeks should be referred.

Personality disorder

BMJ 2013;347:f5276

Linda Gask, professor of primary care psychiatry, Mark Evans, consultant psychiatrist in psychotherapy, David Kessler, senior lecturer

- 4-12% of adults have a diagnosis of personality disorder.
- GPs carry the responsibility for patients with personality disorder (which doesn't mean that GPs are responsible for patients getting personality disorders, although that might be true sometimes), and this is a long term challenge.
- "Personality" refers to the pattern of thoughts, feelings and behaviour that makes each of us the individuals we are.
- People with personality disorders (sometimes referred to as "patients") (but sometimes also referred to as "GPs") experience a persistent pervasive abnormality in social relationships and social functioning.
- "There seems to be an enduring pattern of perceiving, relating to, and thinking about the outside world and the self that is inflexible, deviates markedly from cultural expectations, and is exhibited in a wide range of social and personal contexts "
- People with borderline personality disorder have good remission rates over long periods but recurrence is common, with most showing persistent impairment of social functioning even after specialist treatment.
- Other mental health problems are more common and difficult to treat in people with personality disorders.
- Personality disorders are associated with high use of medical services, medical morbidity and suicidal or violent behaviour.
- Categories include paranoid, borderline, or antisocial personality disorder. Considerable overlap exists between categories. The ICD (International Classification of Diseases) uses a dimensional approach with 5 levels of severity.

- Across the range of personality disorders there is little evidence for what treatments help, although there is some evidence with regard to borderline personality disorder.
- People with borderline personality disorder "tend to seek treatment", whereas those in other categories tend to be reluctant to do so.
- When working with people with personality disorder, explore treatments "in an atmosphere of hope and optimism, building a trusting relationship with an open non-judgemental manner".
- For patients with suspected borderline personality disorder, GPs should help by enhancing coping skills and encouraging the patients to focus on current problems in manageable chunks with agreed follow up times.
- Referral to psychiatrist can be useful to establish the diagnosis or when the patient is in crisis. (Psychiatrist, however, will not see patient for 6 months, by which time patient will be either out of crisis or dead.)
- In specialist care, therapy may include "dialectical behaviour therapy", which involves CBT and mindfulness. There are several other specific therapies for borderline personality disorder, including trials of mentalisation therapy, schema focussed therapy and cognitive analytic therapy, all of which have some positive effects.
- Motivation to engage with treatment may be a problem.
- The type of therapy may not be important (ie. nobody knows what works), but the important thing is that management should be consistent, reliable, encouraging of autonomy and sensitive to change.
- There is no clear evidence that drugs help the core borderline personality disorder symptoms - feelings of emptiness, identity disturbance and abandonment.
- Antisocial personality disorder probably needs multiagency networks that "actively involve" service users.
- GPs still need to offer standard medical care to patients with antisocial personality disorder but need to be aware of the risk of poor adherence and drug and alcohol misuse. GPs also need to assess the history and risk of violence and the presence of other mental disorders.

- Evidence for the effectiveness of psychological interventions in antisocial personality disorder is currently lacking. NICE suggests a group based cognitive and behavioural intervention focussed on reduction in antisocial behaviour.
- Therapeutic communities are used, but there is no evidence of their effectiveness.
- Nidotherapy (nest therapy) is a new approach which manipulates the environment to create a better fit between the person and their surroundings. One study showed improvement in social functioning. (Persuade your patient with antisocial personality disorder to join the police, the army, the banking profession, or get him to become an MP, and bingo! Suddenly he'll fit right in.)
- Personality disorders put a strain on the doctor-patient relationship, Consistency, clarity and forward planning are all important in managing the relationship.
- The diagnosis of comorbid personality disorder is a possibility in patients who do not respond to treatment or seem particularly difficult to manage.
- Conclusion: nobody really knows how to define personality disorder, and nobody really knows how to treat it either, but loads of people have got it, and you're going to have to deal with some of them. Good luck.

QUIZ NO 9

DPP-4 inhibitors work by increasing the breakdown of GLP-1.	☐ True ☐ Fales
DPP-4 inhibitors cause weight gain and are more likely to cause hypolgycaemia than other oral hypoglycaemic agents.	☐ True ☐ Fales
They reduce cardiovascular risk factors by making people stop smoking and take up kick-boxing.	☐ True ☐ False
NICE recommends DPP-4 inhibitors as second line or third line treatment in certain circumstances.	☐ True ☐ False
Pregnant women and children with rashes: Pregnant women are at risk if exposed to children with rashes due to measles, rubella, parvovirus B19 and varicella zoster.	☐ True ☐ False
Measles immunoglobulin must be administered to non-immune pregnant women, who have been in the same room as a child with measles for 15 minutes, within 3 days.	☐ True ☐ False
NAT testing can be done on saliva throat swab or crusted varicella zoster lesions.	☐ True ☐ False
GNAT testing happens in Scotland, during the summer months. It's a way of testing whether you really want to be there or not.	☐ True ☐ False
HNIG (human normal immunoglobulin) may prevent maternal measles.	☐ True ☐ False
VZIG may prevent severe maternal chickenpox and maternal death.	☐ True ☐ False
Headaches: Overuse of opioids, triptans, paracetamol or cheap Australian wine may cause headaches.	☐ True ☐ False
NSAIDs and triptans are best not to be used together for acute migraine attacks.	☐ True ☐ False
Pizotifen and gabapentin are first choice prophylactic treatment for migraines.	☐ True ☐ False

Acupuncture is recommended for tension headaches and migraine.	☐ True ☐ False
Riboflavin is the stuff they use to flavour Ribena.	☐ True ☐ False
Riboflavin is dangerous.	☐ True ☐ False
Oxygen therapy can be prescribed for patients with cluster headaches.	☐ True ☐ False
Patients with medication overuse headaches are always gratified to hear that they've got to stop the medication and there's only a 50% chance that this will cure the headaches.	☐ True ☐ False
If a woman comes to see you with a headache she calls it a migraine.	☐ True ☐ False
If a man comes to see you with a headache he calls it a headache, but thinks it's a brain tumour.	☐ True ☐ False
Telemonitoring and COPD: In the Lothian study, patients randomised to telemonitoring had much fewer admissions for COPD.	☐ True ☐ False
Patients with milder COPD and higher depression scores actually had higher admission rates with telemonitoring.	☐ True ☐ False
Telemonitoring may not reduce admissions, but at least it increases the workload of GPs, and it's about time those lazy fatsos did a bit more work.	☐ True ☐ False
Palm oil: Palm oil is the world's most consumed oil, and it's not very good for your lipid levels.	☐ True ☐ False
Food manufacturers like it because it's cheap, tastes yummy and increases the shelf-life of processed foods.	☐ True ☐ False
Trans fats are even better for you than olive oil.	☐ True ☐ False
Palm oil production has led to the loss of millions of acres of forest in Indonesa and Malaysia.	☐ True ☐ False

A "vanaspati" is a devilish slow ball once bowled by the Nawab of Pathudi. It bowled Geoff Boycott and John Edrich both at the same time, even though they were at opposite ends of the pitch.	☐ True ☐ False
Tamsulosin and hypotension: Tamsulosin is incredibly safe and has never ever led to a hospital admission.	☐ True ☐ False
It would be wise to buy shares in Tamsulosin.	☐ True ☐ False
For safety reasons, Tamsulosin cannot be bought over the counter in the UK.	☐ True ☐ False
Tadalafil has recently been approved for treatment of BPH. Hoorah!	☐ True ☐ False
The combination of tamsulosin and PDE-5 inhibitors has been thoroughly studied in large trials over long periods of time, and every one of these studies has been published.	☐ True ☐ False
Secondary prevention of myocardial infarction: Shorter hospital stays following myocardial infarction have increased the GP's workload and responsibility.	☐ True ☐ False
To compensate GPs for the extra workload involved in cardiac rehab, they've been given a big pay-rise.	☐ True ☐ False
It's a good idea to have cardiac rehabilitation, especially if you are over 95 years old.	☐ True ☐ False
MI patients should have cardiac rehabilitation started within 10 days of discharge.	☐ True ☐ False
The "Mediterranean diet" consists of spaghetti Bolognese smothered in cheese, pizza smothered in even more cheese, and Neapolitan ice cream smothered in chocolate sauce. Yum!	☐ True ☐ False
GPs should estimate left ventricular dysfunction by just looking at patients, and then start them on spironolactone 3 days post-MI.	☐ True ☐ False
The "virtual echocardiogram" technique mentioned above should be taught to all GP trainees.	☐ True ☐ False

Tricagrelor once took on Godzilla in one of those Japanese monster-movies.	☐ True ☐ False
It's completely straightforward for GPs to up-titrate ace inhibitors and Beta-blockers within 4-6 weeks, remember to stop Clopidogrel in 12 months, and stop beta-blockers in (at least) 12 months, except in cases of left ventricular dysfunction.	☐ True ☐ False
The above protocol is made even more straightforward because none of the patients are old or muddle-headed.	☐ True ☐ False
Idiopathic pulmonary fibrosis: Idiopathic pulmonary fibrosis has a median survival of 3 years after diagnosis.	☐ True ☐ False
A key clinical finding is "velcro"-like crackles over the lower lungs and axillae.	☐ True ☐ False
Causes of pulmonary fibrosis include nitrofurantoin and amiodarone.	☐ True ☐ False
N-acetyl cysteine and nintedanib both reduce the rate of decline in lung function.	☐ True ☐ False
It's very hard not to call nintedanib "nine ten and dip", which sounds like a dance move.	☐ True ☐ False
Co-trimoxazole improves quality of life and reduces mortality for idiopathic pulmonary fibrosis sufferers.	☐ True ☐ False
Perfenidone reduces the risk of disease progression by 30% and is recommended by NICE.	☐ True ☐ False
Neck lump: Congenital neck lumps never enlarge.	☐ True ☐ False
Ulcerative skin lesions overlying a lymph node may suggest mycobacterium avium-intracellulare (or scrofula).	☐ True ☐ False
Any new neck lump persisting beyond 6 weeks should be referred.	☐ True ☐ False
Personality disorder: About 1:8 people have a personality disorder.	☐ True ☐ False

Personality disorders cause abnormal social relationships.	☐ True ☐ False
Personality disorders cause inflexible perceptions about the outside world which deviate from cultural expectations.	☐ True ☐ False
It is not unusual for people with personality disorders to have other mental health problems.	☐ True ☐ False
There are lots of categories of personality disorders but there is considerable overlap between categories.	☐ True ☐ False
GPs working with patients with personality disorder should create an atmosphere of hope and trust and be non-judgemental and open in manner.	☐ True ☐ False
If your patient doesn't have a personality disorder, of course, you can be as pessimistic, suspicious, judgemental and deceitful as you like.	☐ True ☐ False
Mindfulness and CBT may be used in people with borderline personality disorder.	☐ True ☐ False
Drugs are particularly useful in treating core symptoms such as feelings of emptiness, identity disturbance and abandonment.	☐ True ☐ False
Effective psychological therapies are available for antisocial personality disorder.	☐ True ☐ False
Nidotherapy involves asking people with antisocial personality disorder to live in nests.	☐ True ☐ False
Patients who do not respond to treatment or who are particularly difficult to mange may have personality disorder.	☐ True ☐ False

Target cardiovascular risk rather than cholesterol concentration

BMJ 2013;347:f7110

Harlan M Krumholz, Harold H Hines Jr professor of medicine

- The recent US guideline on cholesterol treatment and cardiovascular risk focusses on abandoning cholesterol targets. Lifestyle interventions should be tried before drugs are started on the "basis of [individual] patient risk".
- This is a [welcome?] departure from the European Society of Cardiology guidelines which focus on LDL-C targets.
- Modification of cholesterol by drugs "does not always produce the expected result".
- Drugs that lower cholesterol also have adverse effects. The net effect may be an increased risk.
- Drug effects of lipids are a "surrogate effect for outcomes".
- Lipid targets are not based on trials.
- Targets for lipids have resulted in treatment strategies "that have not been shown to reduce patient risk". These treatment strategies have "generated billions of dollars in sales" with uncertain patient outcomes.
- Statins reduce risk of heart disease regardless of initial LDL concentrations.
- The US guidelines "draw attention to absolute risk and the size of the potential benefit, an important point given that statins have tended to be used preferentially in lower risk patients".
- "Risk thresholds for treatment should be understood as recommendations and not dictums".
- Patients should be informed about the risks and benefits of statin treatment, and their feelings should be taken into account.
- It is the doctor's responsibility to provide patients with the appropriate evidence to make an informed choice.
- Guidelines should, perhaps, move more slowly into "performance measures" [?QOF?].

- We should "move past the idea that patients and doctors should defer to thresholds for treatment and develop tools that will facilitate shared decisions about prevention".

Too Much Medicine: from evidence to action

BMJ 2013;347:f7141

Ray Moynihan, senior research fellow, Carl Heneghan, professor of evidence based medicin, Fiona Godlee, editor in chief

- The Quebec Medical Association wants to "wind back the harms of too much medicine". They are concerned about "overdiagnosis and overtreatment", and looking for strategies to counter these thought processes.
- Recent surveys in the US suggest that doctors "inform fewer than 1:10 people about the risks of overdiagnosis and overtreatment with cancer screening".
- The BMJ has been running a "Too Much Medicine" series, which highlights the risks of overdiagnosis across a range of conditions, including pulmonary embolism, chronic kidney disease and pre-dementia.
- The first international Preventing Overdiagnosis conference, supported by the BMJ, was held in 2013, and one of the strategic priorities it identified was to "combat perverse incentives that turn too many people into patients unnecessarily".
- "The Preventing Overdiagnosis 2014 conference will take place on 15-17 September at the Centre for Evidence Based Medicine, University of Oxford. Further information is available at www.preventingoverdiagnosis.net."
- Unfortunately, Dr Hairy's abstract, entitled "Pork Pies are better than Statins", has met with a frosty reception. Perhaps a blindfolded taste-testing wasn't the kind of evidence they were looking for.
- Be that as it may, one of the keynote speakers at the conference will be Iona Heath, former president of the Royal College of General Practitioners, who has recently pointed out that overdiagnosis and overtreatment can have serious ethical implications, eg:
  - unnecessary labelling (telling people they've got a disease when actually they're just normal/a bit old/fond of the odd biscuit)

- making "disease definitions" broader, which can divert resources away from patients who really need them, and
- concentrating on "biotechnical" issues rather than "the wider social and economic causes of disease".

- So there you are, there's really nothing the matter with anybody and all medicine is a waste of time.

- Oh no, wait a minute: "Importantly, the emerging science of overdiagnosis is offered with humility, with ready recognition of the many benefits of medical diagnoses and much unmet need for treatment." Bugger it. Just for a moment I thought we could all pack up and go home.

- A warning note: "Evidence is not produced and used in a value-free vacuum... it is generated, disseminated, and sometimes distorted by vested interests—both professional and commercial... As attempts to wind back unnecessary medical excess intensify, some of those vested interests will no doubt fight back hard to defend their turf and their markets."

- So book your place now for the Not Enough Medicine conference in 2015. Dr Hairy will be doing a presentation entitled "Statins: Better for you than Pork Pies", and testing the proposition by dropping first a statin and then a big pork pie onto somebody's head.

Polymyalgia rheumatica

BMJ 2013;347:f6937

Sarah L Mackie, clinical lecturer, Christian D Mallen, professor

- The life-time incidence of polymyalgia rheumatica (PMR) is 2.4% for woman and 1.7% for men.
- Little is known about the pathogenesis of this disease.
- Core features include bilateral shoulder and hip pain with raised ESR and CRP (90%).
- Ultrasound evidence of synovitis and bursitis may be found. PET/CT/MRI may show signs in non-symptomatic areas.
- CRP may be more sensitive diagnostically than ESR (as it is driven by interleukin 6).
- Symptoms may suggest polymyalgia rheumatica but inflammatory markers may be normal (consider rheumatology referral).
- Features of PMR may overlap with other rheumatological disorders.
- PMR and giant cell arteritis (GCA) are linked "38 times more often than might be expected owing to chance".
- Higher doses of prednisolone are needed for GCA than PMR for symptom relief (expert consensus) (ie. basically no definitive evidence).
- Some patients with PMR may later be diagnosed with rheumatoid arthritis. Anti-cyclic citrullinated peptide antibodies are "rarely present in polymyalgia rheumatica".
- PMR has no clear genetic association.
- A primary care diagnosis of PMR can be made in the presence of clinical features, raised inflammatory markers and a rapid complete response to prednisolone.
- PMR usually occurs in over-70s and rarely in under-50s. Diurnal variations in symptoms occur, worse in the morning for at least an hour. There may be a normocystic anaemia.
- Warn patients with PMR of symptoms of GCA (including limb claudication).

- Statins and hypothyroidism may cause myopathy (possibly with a raised creatine kinase).
- ESR > 100 is not typical of PMR – consider myeloma, renal carcinoma or occult infection.
- Lack of response to prednisolone may suggest an alternative diagnosis to PMR.
- Glucocorticoids relieve symptoms but may not modify PMR disease process.
- A typical starting dose of prednisolone for treatment of PMR is 15mg OD. Symptoms typically resolve within about 3 days.
- Prednisolone dose is gradually reduced over a median period of 2 years (with a great variation).
- Those with higher inflammatory markers at onset are at higher risk of relapse (probably) and need more prolonged treatment.
- Initial prednisolone reduction is 2.5mg 2-4 weekly until reaching 10mg a day, when reduction should be by a 1 mg daily dosage reduction per month. Daily dosage reduction may need to be even slower at doses of 3mg – 8mg a day or less.
- Try organising that for a patient who's getting dementia.
- It is unusual for patients with PMR to need more than 10mg a day of prednisolone for more than a year.
- There are no good trials providing evidence for methotrexate use in PMR. The same applies to tumour necrosis factor antagonists, and probably to leflunomide and tocilizumab.
- Bisphosphonates are generally advised.

Importance of clarifying patients' desired role in shared decision making to match their level of engagement with their preferences

BMJ 2013;347:f7066

Mary C Politi, assistant professor, Don S Dizon, director, oncology sexual health clinic, Dominick L Frosch, fellow, Marie D Kuzemchak, research assistant, Anne M Stiggelbout, professor

- Doctors may make (wrong) assumptions that patients with "limited health literacy or low education" and/or older adults may not want to participate in treatment decisions.
- "Shared decision making is a process during which clinicians and patients collaborate to make health decisions, considering both the best available evidence and patients' preferences."
- Patients' preferences are particularly imporatnt when "evidence" does not point to a clear best choice.
- Doctors should support patients in shared decision making by communicating evidence and helping patients construct preferences, ask questions and state concerns.
- A US national study showed that "primary care clinicians did not engage in shared decision making about common preference sensitive decisions".
- "[Doctors'] inferences about patients' preferences are often inaccurate".
- Deliberation (weighing things up) and determination (making a choice) are both part of shared decision making.
- Shared decision making does not imply that doctors and patients must have an equal responsibility for the final decision.
- In a study of 3000 people, 96% said they preferred to be offered choices about their care and asked their preferences. 52% wanted to defer final decisions to their clinicians but still wanted to engage in deliberation about the choice.

- Doctors should inform patients of the "multiple options" and the importance of their preferences in choosing one as the first step in shared decision making.
- Explaining options and their risks and benefits "can answer the questions that patients need to ask to improve decision making".
- Information provided should be understandable and accurate.
- Doctors should not make assumptions about patients' values and preferences.
- "Once patients are informed, they can decide whether they would like more (or less) responsibility for their health decision. This approach can improve patients' satisfaction, understanding, and confidence in their choices, whether or not they defer final decision making to their clinicians."

Resistant hypertension

BMJ 2012;345:e7473

Aung Myat, British Heart Foundation clinical research training fellow, Simon R Redwood, professor of interventional cardiology, Ayesha C Qureshi, specialist registrar in cardiology, John A Spertus, director of cardiovascular education and outcomes research, Bryan Williams, professor of medicine

- Hypertension results in an estimated 7.1 million deaths a year (equivalent to 13% of total worldwide deaths).

- Resistant hypertension is raised blood pressure (> 140/90) despite treatment with at least 3 antihypertensive agents at optimal or best tolerated doses. NICE guidelines suggest ambulatory BP monitoring should be used to confirm resistant hypertension with a level of 135/85 excluding white coat hypertension.

- Prevalence of resistant hypertension is somewhere around 7.6% - 8.9% of the hypertensive population, according to studies in Spain and the US.

- About 20% of hypertensive patients in England have uncontrollable BP despite taking 3 or more antihypertensives.

- People with resistant hypertension may be 50% more likely to have a "cardiovascular event" compared to patients with controlled BP on 3 or fewer drugs (3.8 year median follow-up).

- The precise long term prognosis with resistant hypertension is unknown.

- Some typical characteristics of resistant hypertension: female, black, over 75 years old, love salt, have had poor BP control for a long time, are obese, have diabetes, and already have atherosclerosis, CKD and left ventricular hypertrophy. (Wait a minute, those are my patients...)

- Factors which may result in "pseudo-resistant hypertension": poor concordance to treatment, white-coat effect, calcified arteries (especially in the elderly), and GPs who don't know how to take blood pressure or who have "clinical inertia". (Wait a minute, I'm one of those GPs...)

- Up to 40% of patients with resistant hypertension were found to have white coat effect on ABPM. Patients with white coat effect are likely to have no target organ damage but may have postural hypotension, dizziness or syncope.
- Obese hypertensive patients in Germany (BMI > 40) have about a 5-fold greater probability of needing 3 antihypertensive drugs compared to people with a normal BMI. Their cars, however, are very reliable.
- Heavy alcohol consumption is associated with raised BP, stroke, and overall poor prognosis. So consume light alcohol instead.
- Cocaine and amphetamines, prescribed or other recreational drugs, may contribute to resistant hypertension, but the effects are variable and unpredictable. Stick to the alcohol. At least you know where you are with that.
- It is important to recommend a salt intake of < 6 g/day. (It's also fun to watch your patients' dumbfounded expressions, because they haven't got a clue what that means in terms of actual food. And, let's be honest, neither have you.)
- 5-10% of patients with resistant hypertension have a secondary cause, most commonly hyperaldosteronism (low potassium, low renin) and CKD, renal artery stenosis and obstructive sleep apnoea.
- Target organ damage supports the diagnosis of resistant hypertension.
- Concordance with treatment is an important consideration, as hypertension is largely asymptomatic. Specialist centres may use "directly observed therapy", which presumably means watching the patient insert the pill into the gob, to observe therapeutic response. Other centres test patients' urine to assess compliance.
- Offer lifestyle advice, screen for secondary causes, maximise concordance and set "realistic goals in achieving blood pressure targets".
- The combination of an ACE inhibitor and an angiotensin receptor blocker "is not recommended in resistant hypertension" because it doesn't work and causes side effects.
- Best (but weak) evidence supports the addition of spironolactone 25-50mg OD as the fourth drug if blood potassium is less than or equal to 4.5 mmol/L.

- If BP response to spironolactone is good but it is stopped because of gynaecomastia, amiloride or eplerenone can be substituted. Potassium levels should be monitored within 2 weeks of drug initiation.
- If potassium is greater than 4.5 mmol/L, doubling the dose of an existing thiazide diuretic should be considered.
- The evidence for adding alpha or beta-blockers "remains empirical in nature".
- Methyldopa, clonidine, hydralazine and minoxidil are further options.
- Use of fixed combination drugs may improve adherence to treatment.
- Percutaneous transluminal radiofrequency sympathetic denervation of the renal arteries (that's all one thing) and carotid baroreflex activation (which is another thing) are being evaluated as therapy. The first of the above has been effective in trials for around 2 years. Renal denervation is not a cure, however – patients need to continue medication.

Umbilical hernia

BMJ 2013;347:f4252

L Barreto, clinical research fellow in paediatric surgery, A R Khan, associate specialist in paediatric surgery, M Khanbhai, general practitioner, J L Brain, consultant paediatric surgeon

- Umbilical hernia is a defect in the abdominal wall underlying the umbilicus through which intestine can protrude.
- Spontaneous closure can occur up to five years old.
- They are more common in Africans and Afro-Caribbeans.
- Paraumbilical hernais are above the umbilicus and have no potential for spontaneous closure.
- If the skin over the hernia is paper thin, there is a risk of spontaneous evisceration.
- Umbilical hernias may seem to be getting bigger after birth, due to the increase in amount of protruding intestine through the same size defect. (It's the diameter of the hernia which is important, not the amount which is sticking out through it.)
- The risk of incarceration or strangulation is probably less than 1%.
- In children less than 4 years old with asymptomatic hernias – reassure, as spontaneous closure may occur.
- Any slight excess of skin post-repair reduces spontaneously.

Doctors' health: taking the lifecycle approach

BMJ 2013;347:f7086

Michael Peters, head of Doctors for Doctors Unit, Omar Hasan, vice president, improving health outcomes, Derek Puddester, director of physician health, Antony Garelick, co-director of Mednet, Christopher Holliday, director of population health, Thomas Rapanakis, service coordinator, Amber L Ryan, research associate

- Increasing demands on doctors increase levels of stress related illness.
- Transitions in life are major stressors but "may also be empowering".
- Medical students begin their training full of idealism, having gained a place at medical school through high academic achievement, which gives tham a sense of mastery and control. This may result in an expectation of error-free medical practice, reinforced by a culture of perfectionism. There is an abrupt change on qualification.
- Continuing personal (and professional) transitions and conflicting loyalties may create feelings of guilt and inadequacy.
- The last years of medical practice may be complicated by loss of medical flexibility, vision, dexterity and stamina. (You just keep prescribing the same old stuff, because you can't see well enough to read any of the new guidelines, and if you do attempt to read them you keep accidentally hitting the wrong key and logging out when you're trying to scroll down the page, and in any case you're bound to fall asleep before you get to the end.)
- Practice and self-image may be affected.
- Society's perception of the doctor has changed, as have patient expectations, and doctors can feel that their professional identity is under threat.
- "In the face of medicine's unpredictability, any call for wisdom in addition to technical ability renders doctors vulnerable to the

charge that their decisions are neither transparent nor accountable."

- In other words, for "I did this because I was following the latest guidelines", read "I acted correctly"; but for "I did this because my years of experience suggested that it was the right thing to do", read "I'm an eccentric old duffer and quite possibly incompetent".

- "Medical schools will need to... prepare their students for a world where the spurious notion of invincibility will not help them... This perspective needs to be followed through all the way to retirement."

- It's not really clear how the term "lifecycle" is appropriate to this article, because a cycle suggests a return to the point from which you began, whereas this article suggests that doctors start out as know-it-all swots and end up as nervous wrecks.

Medicalising unhappiness: new classification of depression risks more patients being put on drug treatment from which they will not benefit

BMJ 2013;347:f7140

Christopher Dowrick, professor of primary medical care, Allen Frances, emeritus professor of psychiatry

- Sadness may result from grief or other life events.
- Sadness is not depression.
- The "broad diagnostic label" of major depressive disorder "has resulted in overdiagnosis and overtreatment" - the stigmatisation of sadness.
- The criteria for major depressive disorder "have not changed since 1980". They capture too heterogeneous a population, and are so loose that ordinary sadness can be easily confused with clinical depression.
- The 1980 DSM-III criteria combined "melancholia" (endogenous depression) with "reactive depression" (milder and linked to life events).
- Severity, ratings and subtype ratings "were generally ignored" in clinical practice and research.
- DSM-5 "allows major depressive disorder to be diagnosed just 2 weeks after a bereavement". Grief becomes a mental illness, medicalising normal human experience of loss.
- Patients with "uncomplicated episodes of major depressive disorder" are no more likely to have a further episode within 12 months than those without any history of major depressive disorder. These episodes and non-melancholic episodes may be better understood as "normal intense sadness".
- Bereavement is very different from "recurrent major depressive disorder". Bereavement is not associated with "suicidality".
- "The prevalance of depressive disorders in the community is stable" (about 2% in England in 2007). Meanwhile, "diagnoses

of depression among Medicare beneficiaries doubled between 1992-5 and 2002-05".

- Overdiagnosis is now more common than underdiagnosis. A primary care study involving over 50,000 patients showed more false positives than missed or actual ones (by about 50%).
- 11% of the US population over 11 years old now take an antidepressant.
- In England, antidepressant prescribing increased at over 10% a year between 1998 and 2010.
- The "homogenisation of major depressive disorder" is "a consequence of heavy drug company marketing" and lack of focus on psychological, social and cultural issues (in normal human life).
- GPs and the public are complicit in this process, as patients request treatment for sadness and GPs are "encouraged by clinical guidelines and indicators" to comply.
- Western societies expect the right to happiness and feel a need to restrict negative emotions.
- Antidepressants have little or no effect on mild depression – any effects may be related to publication bias by drug companies.
- The placebo effect of antidepressants is strong in people with mild depressive symptoms, who are more likely to regress towards the mean.
- Medicalisation of grief substitutes a medical ritual for "deep and time-honoured cultural ones".
- Cross-cultural consultations may worsen these problems, "replacing loss with illness".
- Attend to time, support, advice, social networks and psychological interventions in patients with milder symptoms.
- Listening, advice on exercise and problem-solving (where possible) may help.
- Sharing experiences through (eg.) Healthtalkonline (www.healthtalkonline.org) may help.
- Watchful waiting is good.

Flashes, floaters, and a field defect

BMJ 2013;347:f6496

Ashraf A Khan, specialty registrar in ophthalmology, Ross J Kelly, general practitioner with special interest in ophthalmology, Zia I Carrim, consultant in ophthalmology

- Flashes (photopsia) of white light in the temporal visual field are usually caused by the shrinking vitreous tugging on the retina, and may be triggered by eye movements.
- Floaters tend to "move away" when an attempt is make to look at them, and then settle back in their original position. They are caused by strands of vitreous humour, blood or inflammatory debris.
- Flashes and floaters with sudden or gradual loss of vision and an enlarged shadow advancing centrally are consistent with a retinal detachment which begins with a tear in the retina.
- Myopia, cataract surgery or blunt trauma all increase the risk of vitreous detachment.
- Diabetics may develop new floaters as a consequence of vitreous haemorrhage. Sarcoidosis may cause floaters due to inflammatory debris.
- Some medication (eg. chloroquine) can cause photopsias.
- Most retinal detachments begin in the upper quadrants and cause corresponding inferior field defects.
- An asymmetrical red reflex may be due to vitreous haemorrhage or retinal detachment.
- Up to 85% of patients with flashes and floaters will have a simple posterior vitreous detachment, but "dilated fundus examination is mandatory to exclude retinal breaks". This can be done by optometrists.

Varenicline for smoking cessation

BMJ 2012;345:e7547

Mira Harrison-Woolrych, director

- There is an association between varenicline and serious cardiovascular events.
- "Postmarketing surveillance and other research have identified and investigated several safety concerns" - notably psychiatric effects (such as suicide) and the abovementioned cardiovascular events.
- However, increased cardiovascular risk was not confirmed by a meta-analysis, and "debate about a causal effect continues".
- A Danish study (Svanstrom et al) compared risk of cardiovascular events of varenicline with bupronion and found not significant difference between the two groups.
- However, this study "may not help us determine whether varenicline should be prescribed to patients at higher risk of cardiovascular disease".
- RCTs show that varenicline is more effective than placebo and "leads to similar abstinence rates to nicotine replacement therapy at 52 weeks".
- "When consulting with patients about smoking cessation it is better to state what we know and acknowledge what we don't".
- "If a patient develops suicidal ideation or unstable angina while taking varenicline it is advisable to stop the drug while investigating further."

An introduction to patient decision aids

BMJ 2013;347:f4147

From Drug and Therapeutics Bulletin

- Decision aids help people to make informed choices about healthcare that take into account their personal values and preferences.
- The aids provide evidence based information about options, risks and uncertainties.
- There is evidence that most patients want to take part in decision making.
- Decision aids are available online, in print or on video, eg. "Helping decide between mastectomy or lumpectomy for early breast cancer" can be found on the NHS Direct website.
- Ideally patients should work through decision aids themselves.
- A Cates plot can illustrate risks of treatment with visual aids (ie. smiley faces) (see www.nntonline.net).
- "Tables that present all the options together, along with headline outcomes, have emerged as a popular format" (examples can be viewed at www.optiongrid.co.uk).
- Patients need to consider their values and "trade-offs" of treatment risks and benefits.
- In trials, decision aids "reduced decisional conflict" and decreased patient passivity in decision making.
- "Evidence appears to show that people are more likely to choose more conservative treatment options" after using decision aids. However, people may opt for less conservative treatment at a later date if the "conservative option does not meet their needs".
- The International Patient Decision Aids Standards (IPDAS: see http://ipdas.ohri.ca/) organisation has developed standards for decision aids, and those assessed are available on the IPDAS website (http://decisionaid.ohri.ca/AZinvent.php). (Bah! They're obviously based in California! Probably all high on drugs.)

- The NHS Direct website has lots of useful decisions aids (bah! I've never previously had a good word to say about NHS Direct!) and others are being developed by NHS RightCare (bah! I've never even heard of NHS RightCare!). The URLs are www.nhsdirect.nhs.uk/en/DecisionAids and www.rightcare.nhs.uk/index.php/shared-decision-making/.

- Some aids relating to drugs are available at the Medicines and Prescribing Centre (part of NICE), but the article didn't give a URL for that one, so you can bloody well Google it.

- Decision aids probably increase consultation times by a median of 2.5 minutes. Bah! I'm liking this less and less.

- "Embedding decision aids in clinical systems is likely to make them more accessible to clinicians". Bah! Those who know how to find things on their clinical systems, that is.

- Decision aids "help improve communication between clinicians and patients".

- Bah!

QUIZ NO 10

Target Cardiovascular Risk: It is important to concentrate on LDL-C targets rather than a patient's overall cardiovascular risk.	☐ True ☐ False
Drugs that lower cholesterol never have side effects, especially not myalgia or "feeling liverish".	☐ True ☐ False
Lipid targets are necessary if patient risk of cardiovascular disease is to be reduced.	☐ True ☐ False
Statins are mainly used for high risk patients.	☐ True ☐ False
Too much medicine: Underdiagnosis and inadequate treatment is a major problem for primary care in the UK.	☐ True ☐ False
Overdiagnosis is highlighted in the following three conditions: pulmonary embolism, CKD, "pre-dementia".	☐ True ☐ False
There are no incentives for transforming patients into people.	☐ True ☐ False
The Preventing Overdiagnosis conferences are funded by all the big drug companies.	☐ True ☐ False
Vested interests are interesting people who wear vests.	☐ True ☐ False
Polymyalgia Rheumatica: The best way to get a clear diagnosis in primary care is to commission lots of fancy scans.	☐ True ☐ False
PMR occurs frequently in people under 50 years old.	☐ True ☐ False
ESR is commonly >100 in PMR.	☐ True ☐ False
Symptoms of PMR usually settle after 3 days of prednisolone 15mg OD.	☐ True ☐ False

There is good evidence for the effectiveness of methotrexate, tumour necrosis factor antagonists, leflunomide and tocilizumab in PMR.	☐ True ☐ False
Shared decision making: Shared decision making is a process by which clinicians direct patients into the local CCG priority pathways.	☐ True ☐ False
Doctors' inferences about patient preferences are usually accurate, eg. elderly patients would rather have a nice pair of slippers than a hip replacement.	☐ True ☐ False
Patients want to be offered choices and allowed to express their preferences, but are happy to defer decision-making to the GP in a good proportion of cases.	☐ True ☐ False
Information to patients should be expressed in the most complicated medical language possible.	☐ True ☐ False
Involving patients in decision making improves their understanding, satisfaction and confidence.	☐ True ☐ False
Resistant Hypertension: Resistant hypertension has a very clear long term prognosis.	☐ True ☐ False
About 10% of hypertensive patients in England have uncontrolled BP despite taking 3 or more antihypertensives.	☐ True ☐ False
Many GPs suffer from "clinical inertia", and the really fat ones may get completely stuck in their swivel chairs.	☐ True ☐ False
Hardly any patients with resistant hypertension are found to have the white coat effect.	☐ True ☐ False
The white coat effect is when the white paint on your bedroom ceiling still looks patchy after three coats, and it really makes your blood pressure go up.	☐ True ☐ False
There is strong evidence to support the use of spironolactone as the "fourth drug" in resistant hypertension treatment.	☐ True ☐ False
Percutaneous transluminal radiofrequency sympathetic	☐ True

denervation of the renal arteries and carotid baroreflex activation are being evaluated as therapy. But nobody understands what they are.	☐ False
Umbilical hernia: Umbilical hernias can close spontaneously up to 5 years old.	☐ True ☐ False
It's perfectly OK if the skin is paper thin over the umbilical hernia.	☐ True ☐ False
"The risk of incarceration or strangulation is about 1%." Does this statement apply to inguinal hernias or rogue GPs?	☐ The hernias ☐ The GPs ☐ Both
Doctors' Health: All doctors thrive on stress.	☐ True ☐ False
All newly-qualified doctors are idealistic perfectionists, and the real world comes as a nasty shock to them.	☐ True ☐ False
Older doctors mature and develop like a good cheddar, a fine Port, or possibly a compost heap.	☐ True ☐ False
The problem with wisdom is, it's not transparent. No man can see to the bottom of the dark pool where the magic fish lives.	☐ True ☐ False
Medicalising unhappiness: The DSM diagnostic criteria have resulted in overdiagnosis and overtreatment of depression.	☐ True ☐ False
Sadness is not depression.	☐ True ☐ False
Severity ratings are always used in research into depression.	☐ True ☐ False
If people are sad 2 weeks after a bereavement, they probably need antidepressants.	☐ True ☐ False
Patients with a non-melancholic episode of depression are at greater risk of a further episode within 12 months than patients with no history of depression.	☐ True ☐ False

Depression is being underdiagnosed in primary care.	☐ True ☐ False
It is right for patients to expect medical treatment for sadness.	☐ True ☐ False
Time, support, advice on exercise and psychological interventions can be offered to patients with milder symptoms.	☐ True ☐ False
Flashes and floaters: Flashes and floaters should not be confused with floaters and flushes. One is a problem with the eye, the other a problem with the toilet.	☐ True ☐ False
Myopia, cataract surgery or blunt trauma may increase the risk of vitreous detachment.	☐ True ☐ False
Chloroquine can cause floaters.	☐ True ☐ False
Retinal detachments mostly cause inferior field defects.	☐ True ☐ False
Varenicline: If a patient feels suicidal or develops unstable angina whilst taking varenicline, you should just tell them it's safe to continue, because there are no proven causal links with those conditions.	☐ True ☐ False
If the Eastern European drug Cytisine were licensed for smoking prevention over here, nobody would have to use Varenicline in the first place.	☐ True ☐ False
Patient decision aids: Lots of decision aids are available online, but it might be more useful to have them embedded in your clinical system.	☐ True ☐ False
Decision aids may increase consultation-times significantly.	☐ True ☐ False
I'm all in favour of involving patients in decision making, but I can't stand those bloody patronising grids full of smiley and sad faces.	☐ True ☐ False

Management of urinary incontinence in women: summary of updated NICE guidance

BMJ 2013;347:f5170

Antony Smith, chair of the Guideline Development Group, David Bevan, project manager, Hannah Rose Douglas, associate director, health economics, David James, clinical codirector

- Up to 40% of women going to a "primary care clinic" have urinary incontinence (but not necessarily while they're there).
- New treatments include botulinum toxin A and surgical tapes.
- Categorise the patient's problem "as stress urinary incontinence, mixed urinary incontinence, or urgency urinary incontinence (overactive bladder)" and start initial treatment on this basis.
- Digital examination is recommended to "confirm pelvic floor muscle contraction".
- Bladder diaries are recommended for at least 3 days – covering fluid intake, voiding times/volumes, leaking episodes, pad use and degree of urgency incontinence.
- "Absorbent products, hand held urinals, and toileting aids" should be used as "coping strategy", adjuncts to therapy, or for long term management after other options have been tried, rather than regarded as a form of treatment in their own right.
- 3 months of pelvic floor exercises should be offered for stress or mixed incontinence.
- Duloxetine may be offered as second line treatment for women with stress incontinence, in preference to surgery.
- The full benefits of antimuscarinics for overactive bladder may only be apparent after 4 weeks. They should be started at the lower recommended dose.
- Do not offer oxybutinin (immediate release) to "frail older women" as it "can affect cognitive functioning".
- Oxybutinin (immediate release), tolterodine (immediate release) and darifenacin (once daily preparation) can be offered as first

line for overactive bladder or mixed incontinence. Transdermal overactive bladder drugs can be used if oral medication is not tolerated.

- Consider referral for women who do not respond to drugs and wish to consider other options.
- "Invasive therapy" may be offered for overactivity.
- Botulinum toxin A has only recently gained UK authorisation as treatment for overactive bladder. It may result in the need for self-catheterisation for "variable lengths of time" and increases the risk of urine infections. There is lack of evidence for long-term effects.
- Prompt review is advised if symptoms recur after botulinum treatment.
- Percutaneous sacral nerve stimulation may be used for overactive bladder for women not responding to other treatments. There is a risk of failure and it can cause implant site pain or require surgical revision.
- Immediate release oxybutynin remains first line treatment for overactive bladder as it is the most effective and also the most cost-effective.

Treatment decision aids are unlikely to cut healthcare costs

BMJ 2014;348:g1172

Steven J Katz, professor

- Decisions about testing strategies and disease management are increasingly complex.
- "Interest in decision aids is motivated by expectations that they can increase the efficiency and effectiveness of decision making regarding treatment and improve patients' experiences."
- Decision aids can increase a patient's knowledge, satisfaction and engagement with the clinical encounter.
- It has been suggested that decision aids could reduce costs, with patients opting for less extensive treatment.
- Walsh & colleagues, in their study, "underscore the gaps in the literature" concerning the possibility that decision aids save money. Few articles have dealt with this issue, and they vary markedly in design.
- Walsh et al concluded there was not enough evidence that decision aids lowered costs.
- A recent Cochrane meta-analysis suggested that decision aids can reduce overtreatment.
- The factors contributing to rising healthcare costs include the "increasing intensity of technology and services per capita – largely controlled by clinicians".
- "Proposed strategies to improve quality and value in healthcare are mainly directed at the clinician and system level, including payment reform." Presumably this means making clinicians more responsible for their costs, as a means of discouraging them from further increasing the "intensity of technology and services per capita".
- "Patient decision support tools are best viewed as cost effective rather than cost saving technologies."

- "There is no compelling evidence that decision aids change patient behaviour [at all]... let alone in the direction of less costly treatments."
- There is evidence that patients often have unrealistic expectations regarding the perceived benefits of treatment, therefore "selling decision aids as cost saving technology could do more harm than good if expectations are too high and outcomes fall short".
- Questions arising include
  - What are valid measures of patient engagement in decision making?
  - What are potential candidates for communication-based quality indicators of physician practice?
  - How do family and friends influence decision making?
  - Can GPs really act as effective purseholders for the NHS whilst at the same time pursuing the grail of greater patient involvement in the decision-making process, if it is demonstrated that patients, when more fully involved, tend to opt for more treatment (and therefore more cost) rather than less?

Cutting household ventilation to improve energy efficiency

BMJ 2014;348:f7713

Alistair Woodward, professor of epidemiology and biostatistics

- Global emissions would need to be reduced by half by 2050 to hold warming to less than 2°C (the threshold for dangerous climate change).

- The housing sector contributes about a quarter of national greenhouse emissions, energy efficiency is low and adequate insulation can improve building performance quickly.

- Milner and colleagues point to one health risk: an increase in radon levels in homes in which ventilation is reduced to control heat loss.

- Radon is a noble gas which rises into buildings through cracks in the foundations. "The radioactive breakdown products of radon are potent carcinogens."

- 90% of radon-attributable lung cancers are estimated to occur in homes with radon concentrations below 200 Bq per m3 (only 1% of homes have higher levels than this).

- Milner and colleagues calculated that reducing uncontrolled ventilation would result in a 50% rise in radon concentrations in English homes. This would result in 278 additional deaths from lung cancer a year.

- The long lag period between radon exposure and lung cancer means the effects of more airtight homes would occur 20-30 years in the future.

- Radon and tobacco increase risk of lung cancer in an additive way and the proportion of adults who smoke has halved in the last 30 years.

- If the prevalence of smoking halved again, the additional deaths attributable to airtight homes would be reduced by 44%.

- What about countries which historically have much more airtight homes than the UK, eg. Scandinavia? Has any research been done to invesigate whether radon-related cancer-rates are much higher in those countries?

- To put the radon thing in perspective, if global warming is not averted (which, by the way, it won't be), then in the next 40 years there may be a 5-10 times increase in "mega-heat-waves", as in 2003, when 70,000 excess deaths occurred.

- "Milner and colleagues' study reminds us that large scale interventions may have unintended harmful consequences. But it also points to opportunities for reducing the risks of climate change in ways that minimise risks to health and may improve it."

- On the other hand, what's probably going to happen, given the way the world works, is that we'll get both the radon-related increases in cancer, and the catastrophic weather-events associated with global warming. Yippee.

Quitting smoking is associated with long term improvements in mood

BMJ 2014;348:g1562

Judith J Prochaska, associate professor of medicine

- Oh yeah? Who says so? Come here and say that, you bastard!
- Ahem.
- Nicotine releases neurotransmitters in the brain which induce pleasure, arousal, mood modulation and reduced anxiety. Ah, fags, lovely fags...
- Ahem.
- Abrupt nicotine withdrawal may increase anxiety and depress mood (mostly in the first 24-48 hours, and usually resolving in 2-4 weeks).
- Taylor and colleagues' review hypothesised that people who stop smoking would report improvement in mental health because they do not suffer the negative effects of acute nicotine withdrawal between cigarettes.
- Compared with smoking, quitting was associated with significant reductions in anxiety and depression, improved mood and quality of life. The effects were similar to those of antidepressant drugs.
- The pattern of findings was consistent, and suggests, at least, that stopping smoking does no harm to long term mood.
- The use of e-cigarettes is rising in the USA and Europe, and they have no benefits in aiding smoking cessation.
- E-cigarettes are associated with high rates of continued tobacco use (dual use).
- Research shows that smokers with depression, schizophrenia and post-traumatic stress disorder can stop smoking "without harming their chances of recovery".
- Is it wrong of me to fancy a glass of brandy and a cigar?

Red flags for back pain

BMJ 2013;347:f7432

**Martin Underwood, director, Warwick Clinical Trials Unit,
Rachelle Buchbinder, professor, Department of Epidemiology
and Preventive Medicine, School of Public Health and Preventive
Medicine, Monash University**

- You can't cure someone's back pain by rubbing his back with a red flag.
- What? Well, what does it mean, then? Oh.
- Ahem.
- Nearly all guidelines on low back pain stress the importance of red flags. Downie and colleagues investigated the accuracy of red flags to screen for fracture and cancer in patients presenting with low back pain.
- "It is worrying that doctors are advised to investigate or refer" on the basis of red flags alone. Most people with back pain and red flags "will not have serious conditions", and investigating or referring them may lead to harm.
- Most studies focussed on individual red flags and only 6 considered combinations of factors that are likely to be "of greater clinical utility".
- Downie and colleagues found that, for fracture, only age, trauma, glucocorticosteroid use and presense of contusion were confirmed risk factors. However, the increased risk of fracture was too low to suggest that these were clinically useful.
- The combination of risk factors for fracture does not "perform substantially better than clinical judgement".
- "Data on red flags for the identification of spinal cancer are even weaker"; the only informative red flag is a previous history of cancer.
- There is insufficient evidence to support the formulaic use of red flags in low back pain to aid decision making about further investigation.

- Familiarity with the clinical picture of serious causes of low back pain and active consideration of these diagnoses may be more useful than red flags.
- A high degree of vigilance is needed to ensure the symptoms of Cauda equina syndrome are not missed in people with low back pain with or without radicular pain.
- In other words, don't pay any attention to red flags in the majority of cases where they don't indicate anything serious, but do pay close attention to them in the minority of cases where they do.
- Have you ever considered the possibility that the majority of people with back pain and red flags may be communist proletarians, whose backs have been bent out of shape by the yoke of Capitalist oppression? Well, have you?

Sodium in drugs and hypertension

BMJ 2013;347:f7321

Pasquale Strazzullo, professor of medicine

- Guidelines recommend reducing salt intake to 5-6 g/day.
- The high salt content of commercially marketed food is the main reason for excessive salt intake.
- George & colleagues' study focusses on the sodium content of medical preparations.
- The study involved UK primary care patients prescribed at least 2 sodium containing drugs or matched standard formulations of the same drug between 1987 and 2010. It followed up 1.3 million patients for 7 years, during which time 61,000 cardio-vascular events were recorded.
- For each cardiovascular event endpoint, cases were significantly more likely than controls to have been taking a sodium containing drug (with a dose-response association).
- It was advised that doctors should prescribe sodium containing drugs with caution and not to patients with hypertension (if possible).
- Sodium containing drugs include effervescent and soluble analgesics, cold and flu preparations, indigestion and antidiarrhoeal preparations, bowel cleansing solutions, diuretics and antihypertensives.
- The last item on that list is particularly ironic.
- Salifying drugs makes them more soluble and increases bioavailability.
- Potassium salts are as soluble as sodium salts and sodium is probably used instead of potassium because – you guessed it – it's cheaper.
- Warnings are put on labels of over the counter drugs in the US if they contain >140mg sodium as the maximum daily dose. In George's study, the drugs contained 2.5g/day of sodium. EU regulations now rule that drugs containing more than 1 mmol of sodium per dose should include a warning on the packet.

Ruling out DVT using the Wells rule and a D-dimer test

BMJ 2014;348:g1637

Alfonso Iorio, associate professor, James D Douketis, professor

- D-dimer testing is an index of thrombin generation.
- A low Wells score (≤1) coupled with a negative D-dimer gives a post-test probability for DVT of less than 2%.
- Geersing and colleagues re-evaluated the Wells rule in typical patients and patient subgroups, pooling individual patient data from 10,000 patients with a mean DVT rate of 20%.
- The findings showed that the Wells rule cannot be used alone.
- The combination of a Wells score ≤1 and a negative D-dimer "unequivocally excludes DVT, with an overall failure rate of approximately 1.2%", avoiding venous ultrasound in about 1 in 3 patients assessed for suspected DVT.
- The clinical prediction guide can be used safely in patients with suspected recurrent DVT and should be avoided in patients with active cancer.

Readmission rates

BMJ 2013;347:f7478

Joseph P Drozda Jr, director of outcomes research

- In the USA financial penalties were imposed on hospitals with high readmission rates for acute myocardial infarction, pneumonia and heart failure.
- It is questionable whether readmission rates are suitable as a quality measure, as they have the potential to decrease access to care and there is an inverse correlation between 30 day readmission for heart failure and mortality rates.
- Dharmarajan & colleagues' study in the US found that hospitals with lowest readmission rates had fewer readmissions for all three diagnoses (myocardial infarction, heart failure and pneumonia).
- Donze & colleagues (another US study) showed that approximately 23% of patients were readmitted within 30 days and that 5 of the most common primary readmission diagnoses were related to one of seven comorbidities. Basically the comorbidities are as likely to cause readmission as the hospital diagnoses at the first admission.
- Therefore, interventions focussed on the principal diagnoses at first admission are unlikely to reduce readmissions.
- The better performing hospitals provide care that is not disease specific and that results in good outcomes for all three conditions studied. It is therefore possible that lower readmission rates may be driven by hospitals' organisational characteristics.
- The factors causing readmission may include:
  - inadequate treatment
  - poor systems for handing over care
  - socioeconomic factors
  - progression of disease
- The balance of harm and benefits associated with using readmission rates in incentive programmes for providers needs

careful consideration. Effects might include distracting providers from other quality improvement efforts, and decreasing access for the sickest patients.

Surgery or radiotherapy for prostate cancer?

BMJ 2014;348:g1580

Abhay Rane, professor of urology

- Prostate cancer accounts for over a quarter of all cancers in men in the UK and 90% of new diagnoses are localised disease.
- Two randomised studies compared surgery with observation and found in favour of surgery, especially in young men and those with intermediate and high risk tumours.
- The only randomised trial comparing surgery with radiotherapy is ongoing, and will not report for some years.
- Sooriakumaran and colleagues compared 34,000 men with prostate cancer who initially had surgery or radiotherapy (15 year followup). They found a survival benefit from surgery, which was greater in younger men and those with intermediate or high risk prostate cancer.
- Men who first underwent radiotherapy were 1.5 to 1.7 times more likely to die from prostate cancer than men who initially had surgery.
- Men with non-localised prostate cancer had non-significant results in favour of radiotherapy.

Beware the lies of patients

BMJ 2014;348:g382

Daniel K Sokol, medical ethicist and barrister, London

- The Hippocratic corpus (from the 4th and 5th centuries BC) advises "keep a watch also on the faults of the patients, which often make them lie about the taking of things prescribed".
- Burgoon's study in the 1990s showed that roughly a third of patients claimed to have lied to their doctors (85% admitted to "concealing or equivocating").
  - Reasons for lying include
  - Maintaining the "sick role"
  - Disability benefits
  - Personal injury litigation
  - Avoiding imprisonment or conscription
- Daniel Sokol's doctoral research "identified ways in which patients feigned ignorance to obtain a second opinion or even to test the knowledge of an unfamiliar doctor".
- There are online groups giving patients tips on ways to get a required drug (mainly opiates).
- Doctors must act in the best interest of their patients and respect their autonomy, but these are not absolute injunctions. Complying with the requests of patients must not undermine clinicians' moral and professional integrity.
- Doctors' actions must, from a medical perspective, be properly arguable. If doctors could not defend their action to a group of peers, then they should turn down the patient's request and suggest a second opinion.
- Complicity in deception may be easier in the short term than a long and awkward discussion, but then doing the right thing is sometimes more onerous than the alternative.
- In real life, every GP has to deal with patients who are dissembling in one way or another. Some pretend to be ill when they're not, others pretend not to be ill when they are, and there are various complicated and difficult-to-interpret positions in-

between. Dealing frankly with these patients ("Do me a favour – you were walking perfectly normally on that leg a minute ago!") is stressful, awkward and potentially disastrous, so you have to develop a strategy of pussyfooting. It's one of the many things about general practice that the GP Curriculum won't teach you.

QUIZ NO 11

Urinary incontinence in women: Pelvic floor exercises should be tried for 3 months for stress incontinence	☐ True ☐ False
Duloxetine can be offered as first line treatment for stress incontinence.	☐ True ☐ False
Oxybutynin is very safe for frail older women and never makes them forget what time Songs of Praise starts.	☐ True ☐ False
Oxybutinin immediate release is one of the first line treatments for overactive bladder.	☐ True ☐ False
No drug with "immediate release" in its name can possibly be a good treatment for incontinence.	☐ True ☐ False
The long term effects of botulinum toxin A for overactive bladder have been thoroughly researched.	☐ True ☐ False
Treatment decisions aids: Decision aids can improve patient satisfaction.	☐ True ☐ False
Decision aids lower health costs.	☐ True ☐ False
Decision aids make patients change their behaviour.	☐ True ☐ False
Decision aids are bad because instead of giving patients advice you now just give them a leaflet or direct them to a website.	☐ True ☐ False
Decision aids are good, because no GP can possibly remember all the latest medical advice, or the risk/benefit ratio of every surgical procedure.	☐ True ☐ False
Decision aids are sometimes thought to save money, but they're just as likely to persuade people to opt for expensive treatments.	☐ True ☐ False
Radon is a noble gas with carcinogenic breakdown products.	☐ True ☐ False

Radon is something you put in your bath after a long day in the garden double-digging your veg patch.	☐ True ☐ False
Reducing ventilation in houses may increase the incidence of radon-related lung cancer.	☐ True ☐ False
Global warming could cause a lot more deaths than radon.	☐ True ☐ False
GPs are good at science, so anybody doing this quiz could easily tell me the atomic number of Radon and name three other noble gases.	☐ True ☐ False
Quitting smoking: Nicotine makes people feel good.	☐ True ☐ False
Mood changes related to quitting smoking usually settle in 2-4 weeks.	☐ True ☐ False
Stopping smoking improves mood and quality of life and is just as good as an antidepressant.	☐ True ☐ False
People use e-cigarettes as a way of quitting. They never use them at the same time as tobacco.	☐ True ☐ False
You're never alone with a Pall Mall.	☐ True ☐ False
Red flags for back pain: Most people with back pain and red flags have serious conditions which require investigation or referral.	☐ True ☐ False
Clinical judgement is just as good as combined risk factors in diagnosing fractures in people with low back pain.	☐ True ☐ False
The only informative red flag for identifying spinal cancer is a previous history of other cancer.	☐ True ☐ False
Knowing something about the serious causes of low back pain is better than obsessing about red flags.	☐ True ☐ False
Sodium in drugs: Commercially marketed food is usually very high in salt.	☐ True ☐ False

A lot of drugs are rather high in sodium.	☐ True ☐ False
A high intake of salt or sodium can increase the risk of cardiovascular events.	☐ True ☐ False
Ironically, antihypertensive drugs and diuretics may contain sodium, and therefore increase the risk of cardiovascular events.	☐ True ☐ False
It's a mystery why drug companies use sodium salts (a bit cheaper) rather than potassium salts (a bit more expensive) in their preparations. It defies explanation.	☐ True ☐ False
Ruling out DVT: A low Wells score and a negative D-dimer totally and unequivocally rules out any chance of a DVT in all probability. But it could happen. But almost certainly not.	☐ True ☐ False
Using the Wells rule and a D-dimer test may reduce the need for venous ultrasound by about 50%.	☐ True ☐ False
The Wells score and D-dimer can be used accurately in patients with active cancer.	☐ True ☐ False
Readmission rates: There is an inverse correlation between readmission for heart failure and mortality rates.	☐ True ☐ False
The better performing hospitals had fewer readmission rates for all 3 diagnoses studied, and this might be something to do with how they are organised.	☐ True ☐ False
Readmissions are usually related to the principle diagnosis at the first admission, and nothing to do with the patient's comorbidities.	☐ True ☐ False
Incentive programmes to reduce readmissions would be completely safe for all patients.	☐ True ☐ False
Sadly, many elderly gentlemen are readmitted every year, as a result of getting their braces caught in the hospital front door as they leave.	☐ True ☐ False
Surgery or radiotherapy for prostate cancer: 50% of new diagnoses of prostate cancer in the UK are localised	☐ True ☐ False

disease.	
Observation is as good as surgery.	☐ True ☐ False
Surgery is better than radiotherapy for localised prostate cancer, especially for younger men and those with intermediate or high risk prostate cancer.	☐ True ☐ False
Radiotherapy may be better than surgery for non-localised prostate cancer.	☐ True ☐ False
This sounds like a good subject for one of those decision aids.	☐ True ☐ False
Beware of the lies of patients: Patients hardly ever lie to the doctor.	☐ True ☐ False
There are online groups giving patients tips on how to get opiates out of doctors.	☐ True ☐ False
It would be grossly unfair to label anyone who asks you for a painkiller as a drug-crazed smackhead who doesn't know right from wrong.	☐ True ☐ False
Given the amount of deception going around, it can sometimes be tricky to know where the best interests of your patients lie.	☐ True ☐ False

Withdrawing performance indicators: retrospective analysis of general practice performance under UK Quality and Outcomes Framework

BMJ 2014;348:g330

Evangelos Kontopantelis, senior research fellow, David Springate, research associate, David Reeves, reader, Darren M Ashcroft, professor, Jose M Valderas, professor, Tim Doran, professor

- Data was analysed from 644 GP practices from 2004-05 to 2011-12 and included over 13 million patients.
- The withdrawn indicators included influenza vaccine (asthma) and lithium monitoring (psychosis), removed in 2006; blood pressure monitoring (for coronary heart disease, diabetes and stroke), cholesterol monitoring (coronary heart disease and diabetes), and blood glucose monitoring (diabetes) withdrawn in 2011.
- Mean performance levels were stable after removing the incentives in both the short and long term.
- For the 2 indicators removed in 2006, levels in 2011-12 were close to 2005-6 levels.
- For 5 of the 6 indicators removed in 2011 there was no significant effects on performance after removal.
- The withdrawn indicators were linked to other outcome measures which remained in the scheme.
- All the practices studied used the same clinical computing system (Vision 3). Clinical system is a predictor of QOF performance, and the "generalisability of findings to all practices might be limited".
- If all the practices in the study were using the same clinical system, perhaps that clinical system just left the indicator-reminders switched on by accident, even though the indicators were no longer earning points.

- Perhaps practices just carry on with some of these indicators because they think they're a good idea, regardless of the points system. Doesn't seem very likely, though, does it?
- Perhaps practices carry on with them because they're so bloody confused about what's in the points system these days that they think they'd better keep checking everything, just in case. That seems a bit more likely.
- Perhaps in the future the government will use this research as justification for withdrawing all the points from all the indicators, but insisting that everybody should keep doing all the work anyway. That seems extremely likely.

Political drive to screen for pre-dementia: not evidence based and ignores the harms of diagnosis

BMJ 2013;347:f5125

David G Le Couteur, professor of geriatric medicine, Jenny Doust, professor of clinical epidemiology, Helen Creasey, dementia specialist, Carol Brayne, professor of public health

- The belief that screening for "pre-dementia" is valuable is "creeping into clinical practice", and screening is now policy in many countries.
- In the UK GPs will be paid £3600 a year for assessing patients over 75 years old and those over 60 in at-risk groups (vascular disease or diabetes) for dementia and cognitive impairment.
- There is "no sound evidence that memory clinics are beneficial"; in fact "there is evidence that they may be no more effective than standard care by general practitioners".
- Cholinesterase inhibitors are promoted as potent and effective treatment despite limited evidence to support this.
- Screening for dementia and mild cognitive impairment does not meet the WHO screening criteria as "there is no evidence for the usefulness of any preventive or curative pharmacological intervention".
- The assumption is that people with mild cognitive impairment or "pre-symptomatic Alzheimer's or dementia" will progress to symptomatic dementia over time.
- Only 5-10% of people with mild cognitive impairment will progress to dementia each year and up to 70% do not progress or may improve with reversal of structural changes in the brain.
- Some studies show that development of dementia is higher in people who don't have symptoms of mild cognitive impairment than those that do.
- If the DSM5 criteria for "minor neurocognitive disorder" are applied, then 16% of the population will be automatically defined as having this "disorder".

- Studies of dementia estimated that if a clinician saw 100 patients with a dementia prevalence of 6%, using current criteria he/she would correctly identify 4 of the six but incorrectly identify dementia in a further 23 people.
- Alzheimer's can only be diagnosed by examining brain tissue and finding plaques and tangles containing amyloid and tan proteins.
- In those over 85 years, the prevalence of Alzheimer's brain pathology is similar in people with and without clinical features of dementia.
- Neuroimaging and cerebrospinal fluid markers are increasingly used in detecting people who may go on to develop dementia.
- There are no large population studies suggesting that the association between biomarkers with dementia or neuropathology is "sufficiently robust to be used in clinical practice".
- 65% of people over 80 years have abnormalities on amyloid imaging and so could be diagnosed as having Alzheimer's or "pre-disease", but the amyloid scanning does not predict cognitive function.
- Early diagnosis of cognitive impairment would allow counselling, education and support, but a recent study showed no benefit from these interventions.
- Medication that impairs cognition could be avoided in "pre-dementia", but "this is simply good practice and does not require screening for dementia".
- The adverse effects of cholinesterase inhibitors include hip fractures, syncope and pacemaker insertion. "One trial suggested increased mortality in people with mild cognitive impairment treated with a cholinesterase inhibitor."
- "Dementia is the illness most feared by people over 55 years old." The diagnosis has emotional and social implications and "may result in suicide or euthanasia". [Euthanasia? Mercy-killing lots of other old people?]
- When someone aged 90 years or more dies, the risk of being demented is around 60%. [Or, to put it another way, when someone aged over 90 dies the risk of dementia goes right down to 0%.]

- The emphasis on early diagnosis (of Alzheimer's) diverts attention and resources "from the current needs of older people, which relate to multimorbidity and palliative care".
- Pharmaceutical companies sponsored a study that called for the UK government to provide financial rewards for increased diagnosis rates and development of the Seven Minute Screen for dementia.
- Nearly half the people who have positive results on screening for cognitive impairment refuse subsequent diagnostic evaluation because of concerns about harms associated with a diagnosis.

Where there's smoke . . .

BMJ 2014;348:g40

Michael Brauer, professor, school of population and public health, G B John Mancini, professor, division of cardiology, department of medicine

- Polluted outdoor air is classified as carcinogenic.
- It is the impacts on cardiovascular disease that are the biggest disease burden attributable to air pollution.
- Air pollution is responsible for 1.2 million deaths a year in China.
- A recent meta-analysis reports the relation between long term exposure to air pollution and myocardial infarction and unstable angina.
- Long term exposure to air pollution and observation of an association with increased acute cardiac events suggests that air pollution can be a trigger for such events.
- Nearly 90% of the world's population live in locations where the WHO guidelines for air pollution are exceeded. In a recent study "the mean PM2.5 [fine particle air pollution] concentration over a five year period in Beijing was more than 10 times the WHO guideline value".
- See also the article below, about climate change and human survival.

Does agomelatine have a place in the treatment of depression?

BMJ 2014;348:g2157

Gilles Ambresin, senior lecturer, honorary research fellow, Jane Gunn, chair of primary care research

- Current guidelines recommend that antidepressants should be considered for treating patients with major depressive disorders and dysthymia.
- The efficacy of antidepressants has been overestimated by publication bias.
- Antidepressants are less effective in those with mild to moderate depression.
- Agomelatine works via the malatonergic system and is being promoted as an alternative to second generation antidepressants.
- A meta-analysis of published studies showed agomelatine was better than placebo and (possibly) better than second generation antidepressants measured on reduction of scores on the Hamilton rating scale.
- However, the size of benefit of agomelatine over second generation antidepressants "looks small and may not be clinically relevant".
- Examination of published studies alone overestimated the efficacy of agomelatine by 20%. Once you add in unpublished studies, including one I've just written on the back of an old beer-mat, you get a very different picture.
- "Agomelatine is no more effective than other agents, and, while it may work slightly better than a placebo, the effect size reported by Taylor and colleagues is well below the threshold for clinical relevance."
- "People taking agomelatine need repeated monitoring of their liver enzymes, which further limits the cost effectiveness of this drug in practice."
- "Agomelatine is appropriate as an alternative second line agent in the pharmacological treatment of severe major depression."

- That's a polite way of saying that agomelatine is rubbish.
- If you're looking for a really effective agent, try James Bond.

Climate change and human survival

BMJ 2014;348:g2351

David McCoy, senior clinical lecturer, Hugh Montgomery, director, Sabaratnam Arulkumaran, president, Fiona Godlee, editor in chief

- The intergovernmental panel on climate change (IPCC) concluded "it is virtually certain that human influence has warmed the global climate system" and that more than half of the increase in surface temperature is "anthropogenic". So in your face, Jeremy Clarkson.

- Further global warming may reduce the availability of food and fresh water, increase extreme weather events, cause a rise in sea level, reduce biodiversity, result in uninhabitable areas and cause human migration, resulting in conflict and violence.

- "Business as usual" will increase CO_2 concentrations from the current level (400 ppm) to 936 ppm by 2100 with a 50:50 chance of a mean temperature rise of $>4°C$, which would be "incompatible with an organised global community" (presumably because everyone would be at the beach drinking Pina Coladas instead of getting on with their work).

- "Countries must set aside differences and work together as a global community for the common good, and in a way that is equitable and sensitive to particular challenges of the poorest countries and most vulnerable communities." Yeah, right, that's going to happen. Now that the BMJ has published an article about it, the change must surely come.

- Barts Health NHS Trust has reduced its energy bill by 43% since 2009.

- Health dividends of more environmentally-friendly lifestyles include more active forms of transport (eg. walking and cycling), less red meat and less air pollution.

- Health professionals should "push their own organisations" to "divest from fossil fuel industries completely" and move to renewable energy.

- Climate change is "the greatest current threat to human health and survival".
- Money is the only force which is likely to change behaviour. We've been having well-meaning exhortations for everybody to take action on climate change since the 1970s, and bugger all has happened. Countries which are producing a surplus of green gas should be paid for it. Countries which are producing a surplus of greenhouse gas should be penalised. Thank you very much and goodnight.

When and how to treat patients who refuse treatment

BMJ 2014;348:g2043

Rosemary A Humphreys, specialist trainee year 4, psychiatry, Robert Lepper, mental health act adviser and policy lead, Timothy R J Nicholson, academic clinical lecturer

- About 30% of acute medical inpatients lack capacity to make decisions about their treatment.

- Patients are entitled to make decisions that clinicians think are unwise, and this right is protected by the MCA (Mental Capacity Act).

- Patients can be treated against their wishes if their decision-making capacity is impaired and the treatment is for a serious condition.

- Impaired capacity could be due to mental illness, intellectual disability or physical illness affecting mental function.

- Common law is applicable in emergency situations when there is insufficient time to assess capacity.

- The MCA covers the patient's best interests, not the protection of others, and applies to those aged 16 and over.

- The MCA allows restraint if it is necessary to prevent harm to the patient, but the restraint must be "proportionate in degree and duration to the likelihood of the person being harmed and the seriousness of the harm".

- When assessing patients who refuse treatment, first determine the urgency of the treatment to see if common law is applicable, then determine if the condition is physical or mental (psychiatric).

- Common law is widely used in the emergency setting because there is rarely time for consent. No documentation is needed. ~~So if you're working in A&E, strap 'em down and shoot 'em full of diazepam.~~ However, the decision-making process should be clearly documented in the patient's notes.

- Generally, the MCA is used to treat physical disorders affecting brain function, and the MHA (Mental Health Act) is used to treat mental (psychiatric) disorders. ~~So if you're working in the Psychiatric Department, have 'em sectioned, then strap 'em down and shoot 'em full of diazepam.~~ It is important not to assume lack of mental capacity because the patient has a history of mental illness.
- When assessing capacity for a specific decision, first establish if the functioning of the mind is impaired. Secondly assess whether the person can
  - understand
  - retain
  - use or weigh information relevant to a decision
  - communicate a choice
- Consult with the next of kin or GP to establish the patient's wishes when they had capacity. Check for advance decisions.
- Capacity is decision specific and time specific.
- Section 5(2) of the MHA allows a "holding power" of 72 hours to cover the time needed to arrange MHA assessment.
- Two clinicians, eg. GP and approved mental health practitioner, can assess for a longer term section such as a section 2 or 3 of the MHA. Emergency treatment cannot be given under section 5.
- Section 2 is for assessment and lasts up to 28 days.
- When patients have an established diagnosis section 3, which lasts up to 6 months, can be considered.
- In rare cases, the MCA can be used to treat a mental disorder.
- The MHA can be used to treat physical disorders that directly cause mental illness (eg. HIV encephalitis or profound hypothyroidism), or which result from mental illness (eg. nasogastric feeding in life threatening anorexia or the physical sequelae of a suicide attempt).

Therapist guided internet delivered cognitive behavioural therapy

BMJ 2014;348:g1977

Erik Hedman, postdoctoral researcher

- Internet CBT programmes may be with or without the guidance of a therapist.
- The patient accesses therapy through a personal account in a secure internet-based treatment platform.
- Treatment is in modules, often 8-15, and these include information needed to change behaviour. The modules mainly consist of text, but can include images and audio/video files.
- Contact with the therapist is through a messaging system.
- The therapist spends about 10 minutes a week on each patient. Evidence suggests that therapist contact is more effective than unguided internet therapy.
- A therapist can treat up to 80 patients simultaneously and geographical separation is not a problem.
- Internet therapy has been developed for anxiety and depression but also tinnitus, IBS, chronic pain and sexual dysfunction. There is strong empirical support for internet therapy for depression, social anxiety disorder and panic disorder.
- Meta-analysis shows no significant difference between face to face CBT and internet CBT with therapist guidance.
- So how come my local CCG is forking out loads of money for a face-to-face CBT service which is already, after about a year, unable to cope with the demand?
- They could be spending that money on great big pork pies for the CCG meetings, instead of those bloody terrible sandwiches!

Trigeminal neuralgia

BMJ 2014;348:g474

Joanna M Zakrzewska, professor of pain medicine in relation to oral medicine, Mark E Linskey, professor of neurological surgery

- Trigeminal neuralgia is rare, episodic facial pain that is unilateral, electric-shock-like, and provoked by light touch.
- 10% of patients will not respond to antiepileptic drugs.
- Rarely, trigeminal neuralgia can be secondary to a brain tumour, multiple sclerosis or vascular anomalies.
- Peak incidence lies between 50-60 years old and women are more at risk than men.
- There may be an association between trigeminal neuralgia and hypertension and stroke.
- Microvascular decompression of the trigeminal nerve root is the most effective and durable treatment for trigeminal neuralgia.
- Microvascular compression of the trigeminal nerve root is associated with trigeminal neuralgia in about 95% of patients.
- Demyelination of the nerve root may result in reinforced electrical excitability or loss of inhibition of the nociceptive system. (No, I'd never heard of it either.)
- The diagnosis is based almost entirely on the history.
- The periods of remission get shorter with time and the attacks often get longer. Attacks may occur between 3 times a day or as many as 70 times a day.
- Some patients have background, lower-intensity pain for 50% of the time.
- Some patients may have conjunctival injection, lacrimation, nasal congestion or rhinorrhoea, eye lid oedema, ptosis, or facial sweating. Patients with autonomic features are less likely to respond to surgery.
- Bilateral pain occurs in only 3% of patients.
- MRI of the brain is used to rule out other causes of pain if the diagnosis is not clear cut. MRI may identify extracranial masses, perineural malignancy, cavernous sinus masses, demyelination,

lacunar infarcts, or cerebellopontine angle mass lesions. (No, I'd never heard of them either.)

- NICE guidelines (never heard of them) state that in primary care only carbamazepine should be used and if this fails then the patient should be referred.
- Carbamazepine is the preferred drug and initially gives 100% pain relief in 70% of patients. Tiredness and poor concentration are the main side effects.
- Oxcarbazepine (nope, never heard of it) has similar efficacy to carbamazepine but fewer side effects.
- Alternative drugs include baclofen, lamotrigine, gabapentin and pregabalin.
- In addition to microvascular decompression, palliative destructive procedures are another surgical option.
- Destructive procedures include radiofrequency lesioning, chemical lesioning or balloon compression.
- Microvascular decompression provides patients with an 80% chance of being pain free with a recurrence rate of 10% over 10-20 years. Destructive procedures have a recurrence rate of 50% after 3-5 years.
- Microvascular decompression has a 3% risk of hearing loss.
- Destructive procedures are associated with trigeminal-vagal nerve reflex which may be important in patients with cardiac disease.
- Opioids have no effect on the pain of trigeminal neuralgia.
- An RCT of 25 patients given an 8% spray of lidocaine as opposed to saline had a significant decrease in pain for four hours.
- What's an RCT? Some kind of traffic accident? A teachers' union? How come I never know what's going on?

QUIZ NO 12

Withdrawing performance indicators: If performance indicators are withdrawn, it is likely that performance in that quality area will drop off.	☐ True ☐ False
If we stop asking diabetic men about the performance of their willies, it's likely that their willies will drop off.	☐ True ☐ False
Screening for pre-dementia: Pre-dementia is otherwise known as "being over 40".	☐ True ☐ False
GPs would probably screen for pre-dementia without any financial incentive, because it's self-evidently worthwhile.	☐ True ☐ False
Cholinestarase inhibitors are a proven cure for pre-dementia and they don't have any side-effects.	☐ True ☐ False
Clinical assessment may lead to overdiagnosis of dementia.	☐ True ☐ False
A brain biopsy is the best test for Alzheimer's.	☐ True ☐ False
Biomarkers can be reliably used to identify people at risk of dementia.	☐ True ☐ False
People with Alzheimer's have plaque on their teeth and tangles in their hair.	☐ True ☐ False
Amyloid scanning is a good way of predicting cognitive function, plus it's cheap.	☐ True ☐ False
Screening for pre-dementia is great.	☐ True ☐ False
Where there's smoke: Polluted air is carcinogenic and there may be an association with cardiovascular disease.	☐ True ☐ False
The best way to avoid air pollution is to stay indoors.	☐ True ☐ False
Therefore the best way to avoid cardiovascular disease is to	☐ True

stay indoors.	☐ False
Air pollution is particularly bad in China, which is why their landscape paintings always look so misty.	☐ True ☐ False
Agomelatine: Antidepressants are less effective in cases of mild to moderate depression.	☐ True ☐ False
Agomelatine may be better than placebo and second generation antidepressants.	☐ True ☐ False
The Hamilton rating scale was used by Lord Nelson to evaluate potential mistresses. And you didn't see him getting depressed.	☐ True ☐ False
Pharmaceutical research consists of publishing studies that show your drug to be useful and lighting up a big cigar with those that don't.	☐ True ☐ False
Climate change: Using fossil fuels has almost certainly warmed up the global climate system.	☐ True ☐ False
As a result of increased environmental awareness, GPs have all stopped driving around everywhere and flying abroad for their holidays.	☐ True ☐ False
Barts Health NHS Trust has shown that you can actually do something about your carbon footprint and save a bit of money too.	☐ True ☐ False
Politicians can't be worried about the effect of environmental issues on floating voters, because they're doing their best to ensure that a lot more of them will be floating in a few years' time.	☐ True ☐ False
When to treat patients who refuse treatment: The MCA allows patients to make unwise decisions.	☐ True ☐ False
Patients can be forced to have treatment if they lack mental capacity.	☐ True ☐ False
Common law is useful if there is no time to assess capacity.	☐ True ☐ False
The MCA applies to all age groups.	☐ True

	☐	False
The MCA can be used to restrain a patient. But you're better off with a nylon rope.	☐	True
	☐	False
The MCA is only used to treat physical disorders, and the MHA is only used to treat psychiatric disorders.	☐	True
	☐	False
Common law and the MCA can be applied without the need for paperwork.	☐	True
	☐	False
Internet CBT: Therapist-guided internet CBT is probably as good as face-to-face CBT and much cheaper.	☐	True
	☐	False
It is mostly text based.	☐	True
	☐	False
It demands a large time input from the therapists.	☐	True
	☐	False
It is only used for psychiatric disorders.	☐	True
	☐	False
Trigeminal neuralgia: 90% of patients with this condition will respond to antiepileptic drugs.	☐	True
	☐	False
Trigeminal neuralgia may be secondary to tumours or MS.	☐	True
	☐	False
Microvascular decompression of the trigeminal nerve is effective and durable treatment.	☐	True
	☐	False
The diagnosis must be confirmed by MRI scan.	☐	True
	☐	False
It usually gets worse with time.	☐	True
	☐	False
GPs are only advised to try carbamazepine and not even to consider a lidocaine spray.	☐	True
	☐	False
Destructive procedures may not be as helpful as microvascular decompression.	☐	True
	☐	False

Opioids are effective for trigeminal neuralgia.	☐ True ☐ False
You won't come across it often, but when you do come across it you'll be able to describe the case to your colleagues and say "You know what it turned out to be? Trigeminal neuralgia!" and they'll be really impressed.	☐ True ☐ False

www.ingramcontent.com/pod-product-compliance
Lightning Source LLC
Chambersburg PA
CBHW031827170526
45157CB00001B/206